MYTHS & TRUTHS ABOUT COYOTES

What You Need to Know About America's Most Misunderstood Predator

By Carol Cartaino

With research assistance from Richard Cartaino

Menasha Ridge Press

Myths & Truths about Coyotes

Published by Menasha Ridge Press
Printed in the United States of America
Distributed by Publishers Group West
First edition, first printing

Cover design by Scott McGrew
Cover photograph © Design Pics, Inc. / Alamy
Text design by Annie Long

Library of Congress Cataloging-in-Publication Data
Cartaino, Carol, 1944–.
Myths and truths about coyotes : what you need to know about America's most misunderstood predator / by Carol Cartaino; with research assistance from Richard Cartaino. — 1st ed.
p. cm.
ISBN-13: 978-0-89732-694-0
ISBN-10: 0-89732-694-6
1. Coyote. 2. Coyote—Control. I. Title.
QL737.C22C375 2011
599.77'25—dc22
2010034771

Menasha Ridge Press
PO Box 43673
Birmingham, Alabama 35243
www.menasharidge.com

Photo by Steve and Dave Maslowski

dedication

May I be dead when all the woods are old
And shaped to patterns of the planners' minds
When great unnatural rows of trees unfold
Their tender foliage to the April winds. . . .

I weep to think these hills where I awoke,
Saw God's great beauty, wonderful and strange,
Will be destroyed, stem and flower and oak
I would rather die than see the change.

—Jesse Stuart

acknowledgments

By the time you are finishing up a book, it can be hard to remember every step of the long journey that took you to this point, and every hand that hoisted you up, or helped you over a rocky spot in the path. I hope I haven't forgotten anyone here.

First, thank you to my publisher, Menasha Ridge Press, for their patience and unflagging faith in this project. Especially Molly Merkle and Holly Cross (who gave me my first chance to hear editorial remarks in a soft Southern accent). Thanks, too, to Jack Heffron, who helped move the book from idea to publishing contract.

My brother Richard provided strong moral support, as well as a great deal of deeply appreciated hands-on help with the fact and illustration search. Clayton Collier-Cartaino served as a sounding board of second opinions, complete with witty remarks that helped put things in perspective at times. Susan Waddell gave very useful input on the draft manuscript. Veterinarian Dr. Robert T. Sharp and Amy Sharp Schneider provided expert feedback and advice on many issues, plus their usual irreplaceable bright-eyed encouragement.

I am grateful, too, to coyote experts such as Stanley D. Gehrt of The Ohio State University, who provided a wealth of hard-won information to mine, and Paul R. Krausman of the University of Montana, who was kind enough to serve as an expert reader. The many fine publications and Web sites of the individual states on the subject of coyotes were also invaluable. Books by other authors on coyotes, listed in the back of this book, were an inspiration and treasure trove as well. Sincere thanks, too, to the many people who endured my endless questioning and interviewing on this subject, and to outdoor writer Tom Cross of Winchester, Ohio.

Illustrations help bring a book to life, and I appreciate each and every one of the photographers and illustrators whose fine work enlivens these pages. Many of these people—and state agencies—did so either at no charge or at a rate that made it possible to include them. Special thanks to Russell Link of the Fish and Wildlife Department of the State of Washington, Elise Able of Fox Wood Wildlife Rescue, Inc., D. J. Hannigan, Gerald Tang, and Steve and Dave Maslowski.

This list would be incomplete without a nod to the band of coyotes to the southeast of my pole barn office, whose evening serenades first got me interested in this subject.

about the authors

Carol Cartaino, a native of New Jersey and graduate of Rutgers University (major in English and minor in biology), has had a lifetime interest in animals and the outdoors. Growing up, she had just about every pet animal known and walked home from the library every week with armloads of books on wild animals. All of her best time as a child was spent in the woods near home or on her grandparents' farm in Western Pennsylvania. As an adult she has spent many happy hours hiking, camping, fishing, kayaking, and engaging in outdoor photography and nature study.

For the past 40 years, Carol has been a professional book editor and writer's collaborator, working on almost every subject imaginable within nonfiction, with a strong emphasis on how-to, self-help, and reference. She has helped many authors—from the editors of *Field & Stream* and *Arizona Highways* to scuba divers, photographers, plastic surgeons, veterinarians, and experts on moonshine and cookery—to produce satisfying books. In her ten years as an editor in the Trade Division of Prentice-Hall, Inc., books on nature and gardening were among her specialties. In the decade that followed she was editor-in-chief of Writer's Digest Books in Cincinnati. In the years since, she has been a freelance editor and book doctor for literary agents, publishers, and individual authors.

Since 1980, Carol has also served as editor and collaborator for best-selling author Don Aslett, whose books have sold a total of more than three million copies.

Carol lives with her son and many pets on a 66-acre farm in Southern Ohio, on which she can continue her nature study and listen to the coyote songs.

Richard S. ("Rick") Cartaino, of Fort Myers, Florida, was born in New York City and raised in New Jersey. His favorite job there was driving a 60-passenger commuter bus from the NJ suburbs into New York City. He moved to Florida in 1969, where he worked in retail management in the restaurant and men's clothing business. After that he went into law enforcement and is now a retired Deputy Sheriff of Lee County. Richard likes to watch and photograph wildlife; other hobbies include boating, fishing, swimming, woodworking, electronics, and gardening coupled with tending to his handsome collection of tropical plants and trees. Richard lives with his wife Marguerite and has three grown children and two grandchildren.

Photo by Steve and Dave Maslowski

introduction

Once just a colorful character in children's cartoons or accounts of the Old West, coyotes are now real-life neighbors to nearly every one of us. Coyotes can be found in every U.S. state except Hawaii, and they make themselves at home in suburbs and cities as well as the countryside.

Adult coyotes are too big to be just cute and cuddly, and they remind us of the wolves we humans went to such trouble to exterminate or remove from our surroundings. Most of us have heard or read stories, or watched TV accounts, of coyote problems across the country, and many of us have heard the yips and howls of a coyote pack somewhere right beyond our sight.

The spread of the intelligent, adaptable, and opportunistic coyote across America is one of the biggest wildlife success stories of recent years—but not everyone is happy about it. Parents worry that their children may be attacked by a coyote in the backyard, a nearby field, or a park. Pet owners have the same concern about their cats and dogs. Hikers and other outdoors enthusiasts who encounter coyotes in the wild wonder if they are at risk. Sheep and cattle farmers, and even fruit and vegetable growers, have long been up in arms about losses to coyotes. Deer hunters fear that the rising number of coyotes has reduced the number of deer in many places.

Wildlife experts have noted that where coyotes proliferate, the graceful and beautiful red and gray foxes and other animals are harder to find now.

As with any topic about which little is known by the average person and much is feared or suspected, bring up the subject of coyotes, and myths and half-truths fly. Dating back to the earliest Native American experiences with them, there are plenty of past legends, tales, and conjectures about *Canis latrans* (the coyote's proper name) to add to all of the modern-day finger-pointing at the species. These myths and misunderstandings are rooted partly in the actual habits and activities of coyotes, and partly in our fear of and fascination with them.

Coyotes are a mystery to most of us, whether we see them as a delightful form of wildlife to enjoy or have darker feelings about them. We know they are around, often hear them, but rarely see them. They are hard to observe, track, hunt, trap, or contain. They are not just elusive, but can run fast, jump, and swim with ease. When our furry or feathered friends are threatened by them, we often feel helpless—we don't really know what to do about it.

In these pages I have tried to bring together all of the most frequently asked questions about these interesting and unsettling predators, and give them sound, solid answers. I have concentrated on the urgent concerns and greatest points of curiosity people have about these animals, and attempted to minimize nonessential information and highly technical side issues. In putting this material together, I read and assimilated most of the books written on the subject, as well as a vast number of popular and scholarly articles and reports, including the most up-to-the-moment research. I interviewed coyote experts, from wildlife managers, researchers, and scientists in different states to hunters and trappers, farmers and ranchers, and veterinarians. I collected coyote encounter and experience anecdotes from all over, and then sorted them down to the most reliable. Where the information available seemed incomplete or contradictory, I tried to track down the expert or experts who could fill in the blanks or provide a more satisfying answer. I poured all of my research skills, persistence, knowledge about and experience with nature, and my own personal concerns about coyotes into producing a book that deflates the myths, illuminates and shares the truths, and delivers a few surprises along the way.

There are radically different points of view on coyotes, and it would probably never be possible to produce a book that fully pleased both sides—those who could be called coyote huggers and those for whom the only good coyote is a dead one. I have done my best to stay in the middle, jiggling and juggling on my feet up there on the fence, attempting to give consideration to both camps, yet produce genuinely useful information. This book is light on rhetoric, such as celebrations of the coyote and philosophical musings about him, because that is not its purpose, and because many authors have already done that far better than I ever could.

One last note for now: This is a book of myths and truths about coyote practical matters and questions. The huge and rich body of Native American legends about the coyote—let's not call them myths—are a whole 'nother book, or more like small library.

Photo by Steve and Dave Maslowski

a quick perspective on coyotes

After a brief introduction to the coyote (a thoroughly North American animal), this chapter explains how the coyote managed to colonize the entire United States in the past century, and some of the remarkable qualities that made this possible.

A hundred and fifty years ago or today, wherever the coyote is or goes, a storm of concern and controversy usually follows. You will find here also a short overview of human-versus-coyote history, including the many different ways we have felt, and still feel, about him. These are often hotly contested and conflicting, but never boring!

The coyote is a member of the dog family, which includes not just our familiar friends the domestic dogs, but their wild relatives, including wolves, foxes, and jackals. The coyote evolved in North America more than a million years ago, as did all of the dog family, and any members of that family that spread elsewhere in the world did so via the land bridge across the Bering Strait that used to exist between Alaska and Russia.

The coyote originated here and stayed here, however, and the rich body of coyote lore and legend developed by different tribes of Native Americans is a testament to their appreciation of the animal. For Native Americans, the coyote was often a clown and a trickster, but it was often more than that, too—an important, almost godlike being, that had a lot to do with the creation of the world and what went on in it afterward.

The coyote was well known in Mexico long before it came to the attention of European explorers in other parts of the continent. Some of the first serious notice coyotes received in the United States was from Meriwether Lewis and William Clark, who had been directed on their expedition to report on any new animals they encountered during their travels. In 1823 Thomas Say, a naturalist who had traveled west with a group called the Long Expedition, gave the coyote its scientific name, *Canis latrans* ("barking dog"), after becoming acquainted with its impressive nighttime repertoire. Early settlers and explorers called the coyote things like "prairie wolf," "brush wolf," and "American jackal." But the name that stuck came from the Aztec or Nahuatl word for it, "coyotl."

In the early days of western settlement the coyote was an animal of the open grasslands and prairies of the central United States, northern Mexico, and southwestern Canada. Hunting rodents, raiding an occasional camp, and feeding on carrion when available, it was just a minor annoyance and/or source of amusement for the trappers, cattlemen, traders, and cavalrymen traveling through the coyote's homeland. But after the larger predators such as wolves and cougars were largely eliminated from the West, and sheep were added to the ruminants on the range, stockmen began to see the coyote as the number one enemy of their undertakings. By 1850 or so, the coyote was facing guns and poisons wherever it turned. By 1915, the federal government itself launched a massive predator-removal campaign, which continues in a reduced form to this day. The coyote was the main target of this effort, and in less than a century, more than 20 million coyotes were trapped, shot, and poisoned.

THE START OF AN EPIC MIGRATION

Perhaps partly to escape all of that persecution on the range, and partly because every new crop of coyote pups eventually has to find its own home and territory, and may travel a hundred miles or more to do so, coyotes started spreading out. By the turn of the 20th century, they had made it west all the way through the Rockies to the Pacific Ocean, north into Alaska (following, it has often been noted, the trail of trash and dead animals left by aspiring gold miners), and south through Mexico into Central America. By this time, too, the wolf and other large predators had been removed from the East as well, and all of the tree-cutting that accompanied the conversion of the original great eastern forests into farmland created a habitat much more to the coyote's liking. Now the East was a patchwork of fields and small woodlots, full of the "edge" habitat coyotes delight in.

Coyotes, starting in the early 1900s, were quick to take advantage of all of this. They moved eastward out of Texas, Arkansas, and Louisiana, to the Gulf Coast, and even up toward the Ohio Valley. Another wave went north to the upper Great Lakes region, where they mated with the remnants of the wolf population there, and then continued on eastward to New York and New England, having now become a somewhat different animal—see page 9. Coyotes kept on moving and filling in those lovely vacant territories—no predator of any size (save man) there to stop them—until by 2010, they had reached Delaware, the last continental state, and completed their conquest of the continent. No coyotes have swum to Hawaii yet, but they have managed to paddle or ice-walk their way to places like Nova Scotia, Newfoundland, and Cape Breton Island, the Elizabeth Islands of Massachusetts, Cape Cod, and more. By now it is clear why the coyote has been called "the most successful colonizing mammal in recent history," with a range now larger than any other wild animal in North America.

THOSE COYOTES HAVE NO RIGHT TO BE HERE!

I've mentioned this myth here because it relates to what we've just learned about the coyote. As Gerry Parker, the research biologist who wrote the landmark volume *Eastern Coyote: The Story of Its Success,* says, "Many of us feel that the coyote has no right here, that somehow it should have remained on the western prairies, or that someone, somehow, should have seen that it did so. There is no one to blame, unless we want to blame our forefathers who pushed back the frontier, opened up the forests, and worked hard to create the materialistic society we so value today." The late Starker Leopold, one of the most distinguished zoologists in the country in his day, seconds the notion that we had a hand in this: "If biologists were asked to devise a plan to produce bigger, smarter, and more widely distributed coyotes, they would no doubt recommend doing more or less what had been done in the coyote wars: remove many predators that competed with the coyote, eliminate some of the coyote's traditional food sources, but then, through our agricultural and waste disposal practices, provide it with new and more varied edibles."

We never had this coyote problem until the state decided to release them. I wish they'd never done that!

This particular statement, or something similar to it, is one of the most persistent and fervently believed myths about coyotes in the eastern half of the country. I've heard it myself any number of times, even from otherwise well-informed and educated people. The fact is—and state agencies across the country would confirm this vehemently—that NO state department of natural resources, or other branch of state government, in any state (or the federal government, for that matter, either), ever released coyotes anywhere.

South Carolina, for example, not only denies coyote releases vigorously, but points out that the state itself has actually prosecuted people for releasing coyotes. In Pennsylvania, persistent rumors of state releases of coyotes apparently started with one coyote pup that a state wildlife officer had ear-tagged and radio-collared, hoping that the pup would help him find the den from which a family of coyotes was attacking a local herd of sheep. The pup lost its collar and was shot by a hunter, who decided that its numbered ear tag was clear evidence of the state stocking the creatures.

Western coyotes *have* been released in parts of the East occasionally, but this has always been the doing of private citizens. In parts of the

South, for instance, coyote pups have been imported at times by hunters. Some people say this is because coyote pups are hard to tell from fox pups, but more likely, the importers knew full well these were coyote pups, and thought that the swift coyotes would be fun to hunt with hounds. There has also been conjecture that some eastern tourists brought coyote pups back from visits to the West (how or where would they get them, one wonders) and released them after discovering they did not make ideal pets. It is also rumored that easterners with tree-farm interests in the North imported coyotes to control snowshoe hares snacking on their evergreens. More confirmed is the fact that some servicemen, such as World War II troops from Oklahoma and Texas, had coyote mascots that were eventually released in the East. Similar releases happened even in later years, after Army troops' adopted animals got too costly to feed or outgrew their charm as pets.

At best, all of the above created only small pockets of coyotes in the East here and there, which were insignificant compared with the animal's own immigration to the East, mile by mile, state by state, as pups dispersed from litters and the wave of coyote migration met no resistance.

A BIGGER . . . AND BETTER? COYOTE: THE EASTERN COYOTE

During their travels and colonization of the country, some coyotes actually underwent a transformation. As the pioneering first coyotes moved through the upper Great Lakes region, and then east (some sources say this happened in northern Minnesota, some Ontario, some northern Michigan, and some even Maine), they met up with the last remaining wolves in that area. Since potential partners were hard to come by for both parties at this point, they mated and had offspring. This was an extraordinary event, since hybrids of any kind are rare in the animal kingdom, never mind fertile hybrids, which these were. Wolves will normally attack and kill coyotes, but these were hard times for both partners. The animals that resulted from this epic crossbreeding turned up in northern New England in the 1930s and 1940s, causing great curiosity and consternation everywhere they appeared. They were larger and heavier than the coyotes of the West, with males averaging 35 to 45 pounds, and a few tipping the scales at 50 or 60 pounds. They had larger skulls, stronger jaws, thicker necks, bigger paws, longer legs, and somewhat broader muzzles. Their feet also didn't sweat, which was one more detail clearly showing that those wolf genes were making themselves manifest. All in all, the animals looked a lot like German shepherd dogs, rather than resembling foxes somewhat, as western coyotes often did.

Their behavior, too, showed some differences. Eastern coyotes are more at home in forested areas than western coyotes, and hunt bigger game such as deer more often and in groups (shades of the wolf pack). They also hunt adult beavers, and these chubby creatures can weigh up to 60 pounds. Eastern coyotes are somewhat more social than western coyotes, more playful as pups, and less aggressive to littermates and mates.

A western coyote. Photo by John H. Williams

Scientists are likely to declare the eastern coyote a new subspecies of coyote, and many make the point that this animal is still very much in the process of evolving. A number of experts, including urban coyote researcher Dr. Jonathan Way, argue that this animal, which has about one-third wolf genes, should really be called a Coywolf.

FLEXI-COYOTE

If there is a single word that explains the stunning success of the coyote—its incredible spread and the way it has managed to survive and thrive despite all of the efforts to destroy it—it would have to be "adaptability."

An eastern coyote. The eastern coyote is a cross between coyotes and eastern or red wolves, and, as can be seen in this photograph, it is more robust in build and different in other ways, as noted in the text. Photo by Jonathan Way, www.easterncoyoteresearch.com

Other ways to say it might be flexibility, versatility, the ability to take a different way, to go with the flow and roll with the punches.

The fact that the coyote is one of the most primitive of the wild dogs alive today might be one reason for this. "Primitive" also means unspecialized, and the less specialized you are, the better your chances of surviving when things around you change. Then, too, the coyote originated on this continent in times (the Pleistocene epoch, more than two million years ago) when there were bigger and scarier things around to be eluded than in Asia, for example, where the gray wolf originated.

As we will learn later in this book, coyotes can live almost anywhere, eat almost anything, and even change their breeding habits and social customs to suit their circumstances. They are also very smart—they keep trying things out, keep experimenting, and keep on learning. To give just a few small examples, coyotes have learned how to roll watermelons toward each other to break them open, to drink from drip irrigation pipes, to den in culverts, to run *toward* the sound of a gun when it means fresh-killed deer, and to tell when people mean them harm, and when they don't.

As retired U.S. Fish and Wildlife Service supervisor Dick Randall put it: "The coyote has adapted in every way *possible* to survive. And he's so far ahead of you . . . it seems ridiculous when you stop and think about it. There [are] millions and millions of dollars, there [are] the latest in four-wheel drives with winches, scope-sighted rifles, the very best, the latest thing in toxicants, helicopters, fixed-wing aircraft, *everything* after one poor, dumb, four-footed animal. And he's made an ass of everybody [who's] been chasing him. He's adapted to everything they've thrown at him, and he still survives pretty well. We're supposedly so superior, and yet we've been unable to cope with this animal. Of course, a lot of the coping never needed to be done in the first place."

Ohio naturalist Karl Devine points out, "Coyotes have adapted to humans better than any wild animal that has ever existed." Writer Catherine Reid, author of *Coyote: Seeking the Hunter in Our Midst,* puts it even more dramatically: "Unlike other animals that have succumbed to the destruction of their habitats by monster earth movers, steamrollers, graders, and cement mixers, the urban coyotes have learned to live among and even profit from the intruders."

All of this is why it's often been said that the only things that would survive a nuclear war are cockroaches . . . and coyotes.

THE POLARIZING PREDATOR

The coyote did not accomplish all of this without making a few friends . . . and enemies. Few animals evoke such extremes of feeling as the coyote does. Between those who hate coyotes and would rub at least most of them out if they could, and those who think the coyote is one of the most beautiful animals around, and the best thing that ever happened to modern-day ecosystems, there is almost no middle ground. Very few people are indifferent to this creature. People are afraid of coyotes, mistrustful of them, angry at them . . . or they admire them, are fascinated by them, and thrilled by them.

Which camp you find yourself in often has a lot to do with whether or not you have anything they might imperil. If you have pets, livestock, or even young children that coyotes may conceivably be a threat to, or a favorite type of wildlife they might possibly diminish, it is hard to enjoy coyotes simply and wholeheartedly for the interesting and attractive addition to our surroundings that they are. The following chapters are meant to provide some useful and eye-opening information, no matter where you may be in the continuum.

A closeup of a coyote hunting at Palos Forest Preserve in Cook County, Illinois. Photo by Brian E. Tang/Tang's Photo Memories

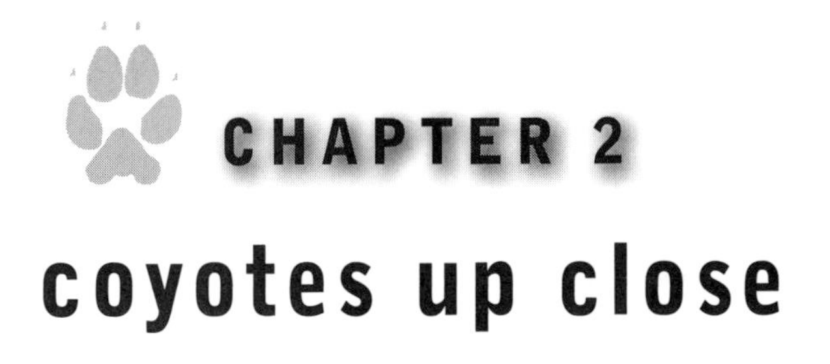

CHAPTER 2
coyotes up close

This chapter takes a closer look at the personality and lifestyle of coyotes, as an aid not only to better understanding them, but also the successful use of coyote control measures outlined in later chapters.

The first question to be answered here is whether or not you actually have coyotes to contend with. This chapter will make very clear what coyotes look like (and do not look like)—how large they are, what color they are, and how they carry themselves. It will also point out the notable differences between western and eastern members of the species; where you are most likely to see coyotes; the surest ways to identify them, and to distinguish them from dogs, wild and domestic, and wolves; and how to tell coyote tracks from dog and other similar tracks. Included, too, is a full description of the coyote diet, since this is where at least 75% of the objections to coyotes originate.

You also will find answers here to questions such as: Just how clever are coyotes in reality? How keen are their senses? Where do coyotes spend most of their time, and why do we see them so rarely? What are their activity patterns, in the daytime and at night, and in different seasons of the year?

EXACTLY WHAT DO COYOTES LOOK LIKE? I MAY HAVE JUST SEEN ONE, BUT I'M NOT SURE . . . .

Probably at least half of the people who live in rural areas and an ever-increasing number of people in the suburbs in the United States have seen a coyote at one time or another, whether they realized it or not. Coyotes have often been mistaken for dogs. At a quick glance, they can be taken for medium-size German shepherds, and their thin, pointed heads and slender frames remind some people of collies (or the more exotic similar-looking breed called the Belgian Tervuren). Interestingly, these dog breeds that coyotes are sometimes confused with are way up there on the canine intelligence scale, as are coyotes themselves.

In general, coyotes are thinner and more graceful and have longer legs than most dog breeds, and when walking or running they carry their tails down, not up, as dogs do. In addition, their big ears—three to four inches high—are always upright, never drooping or curled over like many dogs'.

Coyotes have been mistaken for wolves by people unfamiliar with wildlife, but they are significantly smaller and shorter than wolves, and have thinner faces, frames, and muzzles. Wolves carry their tails horizontally when walking or running, not down like the coyote. Of course, wolves are also rare in this country, and can weigh more than 130 pounds.

Coyotes are actually much closer in appearance to the African black-backed jackal than to any other wild canine in this country. (Check out this animal on the Internet if you have any doubt.) For their similar looks, as well as for their habits in general, coyotes have been called the American jackal. The word or term "jackal," we may note, has almost as many negative connotations as "coyote" has acquired over the years, and for some of the same reasons.

Coyotes close up

The photographs in this book will help you recognize coyotes on sight, and let's zero in on a few specifics of their appearance now.

In general, western coyotes are smaller and more foxlike in appearance than eastern coyotes. And male coyotes are not only larger, but have broader muzzles and foreheads than females. When thinking of coyotes, in short, remember: **upright, pointed ears; thin face; pointy nose; long legs; small feet; slender frame; bushy tail; coyote colors** (described below).

Weight

Coyotes can weigh anywhere from 20 to 50 pounds, and a few have exceeded that amount (the biggest coyote recorded to date was a hefty 74¾ pounds). Everywhere on the continent, males are usually larger than females, and coyotes in the Eastern United States tend to be larger than those in the West (and some extra-large ones reside in parts of the Southeast). The size and weight of coyotes are often overestimated because of their thick fur.

It is hard to generalize within this wide range, but western coyotes tend to weigh on average 20 to 25 pounds, and eastern coyotes from 30 to 45 pounds. Coyote guru Stanley D. Gehrt of The Ohio State University and the Cook County, Illinois, Coyote Project says that most adult coyotes in the areas he has studied weigh between 25 and 35 pounds, and that urban and suburban coyotes are not larger than their country cousins, despite some rumors to that effect.

Height and length

Coyotes are about two feet high at the shoulder, with western coyotes sometimes being a little shorter. Note that wolves can be more than a foot taller than this. In length, coyotes are 40 to 60 inches, with most estimates describing them as "four to five feet" from tip of nose to tip of tail.

What color are coyotes?

The answer to this could fill an entire chapter of this book, if every possible variant were described in detail. Coyotes come in a wide range of color patterns, because they are camouflage artists and tend to match the area where they live. For example, coyotes in mountain regions are usually darker, and desert coyotes lighter. Coyotes of open country are often splotchy gray and white, while those in timbered country have more dark gray and red. But here are some generalities that will help boil this down.

If you were trying to summarize the color of coyotes in a sentence, you would say that most coyotes are gray or tan or some mixture thereof. But coyotes everywhere are actually some combination of the coyote palette of colors: **gray, tan, reddish-brown, black, buff, and white.** Coyotes have a dense double coat with long (usually black) "guard hairs," which often gives them a grizzled appearance, and is one reason they are sometimes mistaken for German shepherds.

The throat, belly, chest, and the fur right above and below a coyote's lips is usually white or buff, and their ears and the base of their ears, cheeks, forelegs, and paws are often reddish-brown.

This photo of an Ohio coyote, even in black and white, shows the rich variations in the color of a coyote's coat. Photo by Steve and Dave Maslowski

Peruse the pages of this book for those pictures that are worth several thousand words.

THE COYOTE'S ARSENAL

The coyote's success, as a predator and in general, has a lot to do with his adaptability, as noted earlier, and also with the abilities that combine to make him, as Professor Stuart Ellins of California State University puts it, "a magnificent hunting machine." First among these talents are his finely honed senses.

The keenest coyote sense: Scenting

The long, skinny coyote nose (with its abundance of scent-processing interior surface) should be a hint that the coyote's sense of smell is his strongest suit. The biggest reason so many would-be coyote hunters come home empty-handed, and prospective trappers are foiled so often, is the coyote's superb scenting ability. Dogs and wild dogs in general have as much as 1,000 times the number of olfactory receptors as we do. Coyotes are exquisitely tuned to odors in the air or on the ground, and if the wind is right they can smell a person more than a mile away. They can hone in on scent clues to prey or carrion from great distances, and follow a prey animal's scent trail right to him. The most successful hunters and trappers take care to disguise or minimize human odor. As coyote hunter Mark Jacklich reported in Jessica Curry's article "The Evolution of the Urban Coyote" in *Chicago Life,* "Coyotes have noses second to none. You can fool their eyes and ears, but you can't fool their nose . . . when a human goes into McDonald's and sniffs the air, he smells a Big Mac. When a coyote smells a Big Mac, it smells the whole thing—two patties, special sauce, cheese, pickles, and a sesame seed bun."

Photo by Gerald D. Tang/Tang's Photo Memories

Coyote talent #2: Hearing

Coyotes have big ears, well suited to gathering sound, and their hearing is extremely sharp. Though different sources credit them with various degrees of extraordinary hearing ability, even more conservative ones say that coyotes can hear far better than humans, and can hear higher frequencies than dogs. Some even feel that coyotes can hear ultrasonic sounds, which helps them in their rodent-hunting. Coyotes can pin down the source of a sound pretty closely from quite a distance, and their acute hearing combined with their sense of smell makes them hard to hunt, capture, or even observe.

How well can coyotes see?

Not that much better than we can, actually, even at night, but they do have better peripheral vision. They can pick up movements at great distances and are quickly aware of anything that looks out of place in their surroundings.

Coyotes, like dogs and cats, do not have cones in their eyes and thus are color-blind—they see the world in shades of black, white, and gray.

Other coyote "senses"

These are not exactly senses in the usual sense of the word, but coyotes have been credited with some other abilities that give them additional help in locating prey and avoiding danger.

One well-known "sense" is their tendency to avoid new things in their environment—the fancy word for this is neophobia, or fear of the new. This is a handy instinct to have when the latest new thing in your surroundings is very likely to be a trap of some kind.

Coyotes are also extremely wary—consider how many places they live where they are seldom seen. They are for the most part well aware that safety lies in not being noticed by people, and they usually rely upon all of their senses, not just one or two, to scope out a situation. Films from a remote wildlife video camera will show, for example, raccoons gustily and happily snacking away on a bait pile, while the coyotes are flitting in and out, tense and nervous-looking, ready to bolt at any moment.

Some people are even convinced that coyotes have a form of ESP—the ability to sense what is going to happen before it happens.

HOW SMART ARE COYOTES?

Coyotes do not appear on any of the most recent lists of "all-time smartest animals," which usually include creatures like great apes, dolphins, elephants, whales, rats, pigs, crows, and believe it or not, parrots and pigeons. But anyone who has ever had direct personal experience with coyotes says that they are VERY smart indeed, and an old Mexican saying insists that they are "the smartest person next to God."

Yet we note that their nickname, beloved of headline writers, is the *wily* coyote. In accord with the reputation that coyotes have acquired through decades and centuries, this may mean intelligent, but with a twist: toward cunning, clever, scheming, crafty, tricky, stealthy. Even foxes, which are famous in lore and legend for their slyness, have a slightly nobler profile. The coyote's wariness, combined with his sharp senses, and his ability to see every possible advantage and take it, have left him "wily" rather than "intelligent."

Coyotes are at least as intelligent as the smartest dogs, and they are better than dogs at learning by observation—one reason it's said that you never fool a coyote twice. Some people feel coyotes are smarter than wolves.

Just a few examples of why many people feel you should never underestimate a coyote:

- They have an uncanny ability to somehow identify events or objects that imperil them, even though it is sometimes hard to imagine how they can do this. Thus, they become very hard to trap in an area once a few have been caught there, will avoid poison baits after a number of them have been poisoned by such a bait, and so on. It is suspected that coyotes have some way of communicating such things to one another. They have even been known to dig up traps and urinate or poop on them.
- People tell of seeing coyotes do things like build earthen dams so that the runoff from an impending rainstorm will flood a gopher burrow and drive the animal out, or stop up prairie dog holes so that when the dog runs for cover he is out of luck. There are even stories of coyotes using the cover of a moving train to sneak up on unsuspecting animals on the other side of the tracks.
- Hunting pairs of coyotes have been observed having one member of the pair engage in antics—such as prancing or rolling—meant to distract the prey, while the other one moves in for the kill. One onlooker witnessed a coyote on a chain that left some of its dog food in sight of hungry free-ranging chickens, within the reach of its chain. Then the coyote lay down nearby and pretended to sleep, until a chicken dinner took the bait.
- Sheep ranchers have reported that when they try to protect their sheep by putting them in buildings at night, the coyotes switch their lamb dinner schedule to 5 p.m.
- Many people swear that coyotes can tell whether or not you are carrying a gun, though this has never actually been proven.

Interestingly enough, when I looked for clear-cut facts and measurements of coyote intelligence, they were not easy to come by. So we might conclude that coyotes are smart enough to convince us all that they are smart, with or without a file cabinet full of evidence.

ARE COYOTES ACTUALLY COWARDLY?

Have you ever noticed how many of the qualities we dislike or find unpleasant in people have been attributed to the coyote? Over the years, coyotes have often been referred to as not just cowardly but "skulking," "slinking," "sneaky," "crafty," "cunning," and the like—not the most flattering ways to say "cautious." Coyotes, as noted elsewhere in this book, are usually exceedingly wary. They were surely somewhat like that originally, in order to elude other, larger animal predators, and more than a century of relentless pursuit and persecution by humans has only intensified this trait. The coyote for the most part is well aware, as noted earlier, that his very survival depends on being noticed as little as possible by man.

D. J. Hannigan, law enforcement, Park County, Colorado

WHERE ARE YOU MOST LIKELY TO FIND THEM?

"Just about anywhere" is the first answer that comes to mind today, but although they can be found everywhere from the desert to farm fields and forests, right up to suburban backyards, coyotes do have their preferences.

When the coyote first came to the attention of white people in what is now the United States in the early 19th century, it was largely a creature of the open grasslands of the West and of the midcontinent. Coyotes are still perhaps best suited to hunting on open ground, so one place they are sure to be found is in grasslands of all kinds, dry and wet. This includes the high, grassy slopes below the timberline in mountainous areas, and the sparsely vegetated, open northern expanses called tundra.

Today, however, the coyote is also very much the master of the "edge"—such as the edges of forests that are near open areas, or hedgerows that adjoin or divide open fields. It loves country that is a patchwork of farm fields and forest. Edges are a great place to travel unobtrusively, and there is usually quite a smorgasbord of prey there. Other wildlife likes edges for the same reason the coyote does—they are a great place from which to dart out and take advantage of some windfall in an adjoining area, and then dart back to the safety of cover.

Coyotes also prefer brushy areas of any kind, from the shrubby openings of forest areas, to thickets and overgrown fields—anyplace with lots of low, dense vegetation, which again provides excellent cover.

Good water sources are very important to coyotes, too, so areas near ponds, creeks, and rivers, including marshes and swamps, gullies, and river and creek bottoms, are another favored habitat. Such areas are also rich in wildlife of other kinds—prey!

Coyotes are not as plentiful in fields filled with farm crops (with the possible exceptions of ones they can feast on, such as corn), as they are in pastures and ranchland.

And although coyotes do frequent forests, they are fonder of open forest and burned or clear-cut areas, than dense, deep forest, or rain forest.

In rural settings, they are not as plentiful in places heavily frequented by humans, as they are in more remote and private places. In suburbs and cities, some of their favorite hangouts are parks and cemeteries.

WHAT DO COYOTES EAT?

An incredible number of different things, and it is their eating habits, in fact, that are responsible for probably the two most important forces in their lives: their ability to succeed almost anywhere, and the fear and hatred they stir up in humans.

A lot is known about what coyotes eat, because in their anxiety to know exactly what that might be, people have made an amazing number of studies of the contents of coyote stomachs. One study back in 1931, for example, examined more than 14,000 coyote stomachs!

Coyotes are technically carnivores, but as eastern coyote expert Jonathan Way has said, in practice they are omnivores. At times, you could mistake them for vegetarians. Their highly varied diet has helped them survive and prosper, and that diet is partly made possible by the teeth they are equipped with. Although most of their teeth are designed to grip and tear meat, like those of most carnivores, coyotes also have molars with large grinding surfaces that enable them to crush and chew up vegetation.

Coyotes will eat almost anything, but they usually concentrate on whatever is most available and abundant in a given area, and easiest to find or catch at that time of the year. Thus their diet often differs quite a bit from season to season.

If you attempted to summarize the endless array of things coyotes eat, and identify their most common main dishes, it would go something like this (the order of these staples might vary in different parts of the country, and according to whether the coyote in question is urban or rural): rabbits and other mammals in that size range, rodents, deer, fruits and vegetables, carrion, and livestock. That last category—see Chapter 4 (page 75)—is not necessarily a large part of their diet.

Those are the general outlines, but the following lists give you a better idea of the diversity and the friction points.

COYOTE FOODS

Wild Foods

Rodents Voles (field mice), mice and rats of many other kinds, shrews, squirrels, both ground squirrels and tree squirrels, gophers, prairie dogs, chipmunks, and more

Rabbits Jackrabbits, snowshoe hares, and cottontails

Big varmints Groundhogs, raccoons, opossums, and skunks

Miscellaneous other wild mammals Muskrats, young beavers and otters, porcupines, weasels, armadillos, javelinas

Deer and similar animals Whitetail and mule deer fawns and adults, pronghorn fawns, occasional elk calves or bighorn sheep kids, occasional elk and even moose

Birds Including songbirds, quail, grouse, pheasants, water birds, and wild turkey poults, and the eggs of ground-nesting birds, including geese

A cottontail rabbit, one of the coyote's favorite foods. Photo by Todd Schneider, Georgia DNR, Wildlife Resources Division

Cold-blooded critters Salamanders, frogs, lizards, snakes (including rattlesnakes), turtles, crayfish, crabs, fish of all kinds, snails

Bugs Large insects like grasshoppers and cicadas, june bugs and other beetles, crickets, centipedes, grubs, worms, bees and honey. Orphaned pups can often survive on insects until they are large enough to catch bigger things, but adults eat big insects with relish, too.

Wild fruits, nuts, and such Including persimmons, wild plums, chokecherries, mulberries, raspberries, blackberries, elderberries, and other wild berries, wild grapes; acorns, beechnuts, pine nuts, and other wild nuts; juniper berries, prickly pear fruits, mesquite pods, and grass

Carrion Coyotes prefer fresh meat but will eat almost any kind of carrion, from dead grizzly bears to dead coyotes or dead humans. The carrion coyotes feast on includes roadkill, animals that die after being wounded by hunters, and the leftovers of other predators such as bears or cougars.

A cotton rat, high on the coyote menu in the South. Photo by Todd Schneider, Georgia DNR, Wildlife Resources Division

Foods Snatched from Humans

Hunters' prizes Deer and other animals killed by hunters and left unattended

Livestock Sheep and lambs, goats and goat kids, newborn calves, young pigs, domestic rabbits, chickens and ducks, cats, small dogs

Fruits Including cultivated berries, cherries, grapes, apricots, peaches, plums, pears, apples, watermelons and cantaloupes, oranges, tangerines, dates, figs

Vegetables Including corn, cucumbers, squash, carrots, tomatoes, peppers, asparagus, potatoes, and onions

Cultivated nuts Plus peanuts

Garbage And all of the goodies therein, from chicken bones to pizza crusts

IS IT TRUE THAT COYOTES ARE BLOODTHIRSTY KILLERS?

Coyotes are predators, and as researcher Stan Gehrt of The Ohio State University has pointed out, we humans began our lives as possible prey, so deep somewhere inside we still have some pretty strong feelings about any large animal that lives by killing others. Most of us are also uncomfortable with the act of killing, so predators of any kind may fascinate us, but they also repel us by the very way in which they have to fill their bellies. (Microscopic or vegetable predators, no matter how deadly, don't usually stir us up to the same degree.)

On the other hand, the mammals called predators are only doing what they are genetically programmed to do, and for survival, at that—they do not sit up nights (like human predators) planning nasty ways to end the lives of others. Researchers have also noted that when predators kill prey, they do so in a very matter of fact, dispassionate way—no gloating or taking pleasure in the act here. They are just doing what they have to do to stay alive themselves.

And then there is the question of just how hypocritical we are in condemning animal predators, as noted by many, with such comments as: "Everything that lives preys on something else; and the most wanton, ruthless, mindlessly destructive predator of all is man" (Pat Wheat of Boulder, Colorado). And, "Civilized man, filled with lust to kill and with morbid righteousness against any other animal that kills" (J. Frank Dobie, author of the great coyote classic *The Voice of the Coyote*). When many of us humans hunt today, we are doing so for something close to recreation; most of us are overnourished already.

HOW FAST CAN A COYOTE RUN?

Very fast. In fact, coyotes are among the top ten fastest animals alive, although they are near the bottom of that list. People who have chased coyotes with cars have clocked the animals' speed at more than 40 miles per hour (many sources say 43 miles per hour), for distances of up to a couple of miles. That puts the coyote below famous speedsters like the cheetah, pronghorn, lion, and quarter horse, but it is fast for a predator. The fastest dog, the greyhound, can only reach 45 to 48 miles per hour.

At their fastest, coyotes can run almost as fast as one of their fleet-footed main prey species, jackrabbits. Jackrabbits can run 35 to 45 miles per hour, hop up to 50 miles per hour, and jump up to 20 feet. Whitetail and mule deer, two other favorite prey, can run from 30 to 40 miles per hour. They can outrun most dogs—and coyotes—in their initial sprint, but often lose the race over time and distance.

More normal speeds for running coyotes are probably 25 to 35 miles an hour, especially in rough terrain, and their most typical mode of travel is the trot. Coyotes are tireless travelers, and can move at a fast trot—their general ground-covering and investigative mode—for hours.

Coyotes are good jumpers, too, capable of leaping 12 or 13 feet in a single bound. And even in the driest parts of the country, coyotes are strong swimmers. They are not, however, great climbers (we should probably be grateful there is something they do not excel at!).

Coyotes are terrific athletes—strong, fast, and agile—and when you combine these traits with their intelligence and learning ability, it adds up to one very formidable predator.

WHAT DO COYOTES ACTUALLY DO ALL DAY? AM I MORE LIKELY TO SEE THEM IN THE DAYTIME OR AT NIGHT?

To answer this question, we have to take "day" in its larger sense—that is, a 24-hour period. In their original prairie homes before European settlers arrived in this country, coyotes traveled about a great deal in the daytime. They still do, in undeveloped areas, wildlands, and anywhere they feel comfortable and unthreatened. And we may see them abroad in daylight almost anywhere, even now, during the pup-raising season, when both parents have to do a great deal of hunting to keep all of those little mouths fed. Young coyotes, such as half-grown pups, are very likely to be active during the day, as well.

However, for the most part, the more people who inhabit an area, the more nocturnal coyotes are there. Thus urban and suburban coyotes are more nocturnal than rural ones, though they may alter this pattern to take advantage of special circumstances they become aware of, such as someone faithfully pouring out big bowls of dog or cat food during the day.

In summer, coyotes are even more likely to be night hunters, since hot midday is not the best time to be moving around in a warm fur coat, and at this time of year, prey and fruits and vegetables, too, are plentiful. In winter, when food is much harder to come by, coyotes may be forced to do some daytime hunting just to stay alive.

Sunrise, sunset . . .

Even when coyotes are largely nocturnal, their all-time most active periods are when day eases into night, and vice versa. This means twilight and daybreak. As many of their prey species—many of them nocturnal too—

are setting out for their night's hunting or grazing, or winding down before the next day's rest, the coyotes are out there, paying close attention.

The ghost on the move

In general, whenever human activity slows or stops, coyotes creep out and do their thing. Except for the brief period each year devoted to mating, this largely means prospecting for food. In one form or another, this is what occupies most of a coyote's waking hours. In the nightly (or daily) quest to fill their bellies, coyotes follow game trails, livestock trails, human trails, dirt roads, other roads, train tracks, and power-line corridors. Just like us, they prefer the path of least resistance, whatever route will take them where they are headed the most direct and easiest way. They usually take the "back way" into places, traveling in the spots few people look at or pay attention to, such as the edges or back of things and in gullies, washes, and the like. If you have ever taken a

Coyotes cover a great deal of ground every day and night in search of food, often at a trot. Photo by D. J. Hannigan, law enforcement, Park County, Colorado

train trip, you have seen how many parts of our everyday environment we are scarcely aware of, but they are there. Urban coyotes, especially, cross many a road in their forays, and the speed with which they do this can be awe-inspiring.

When just exploring for possibilities, they move at a trot, or quick shuffle. When they feel safe from observation or interference, they dart into farm fields and backyards, survey picnic areas and roadsides, many a place they would never venture to in the full light of day.

Naptime

When the time comes for a coyote to sleep, he often does so at midday, or the hottest part of a day. His "bedroom" is some protected area, usually one with good visibility. He doesn't need an actual shelter of any kind, even in the winter, just some thick brush, a fencerow, thicket, pile of rocks or uprooted trees, or field of tall grass or weeds, or grain, that he can curl

A coyote in Colorado doing the traditional coyote leap-and-pounce, mouse-hunting in the snow.

up amidst. In urban areas, he may even consider an abandoned building. But the ever-cautious coyote, even in captivity, usually prefers not to be enclosed in any way during his naptimes. Coyotes only use burrows, holes, or caves in denning time, or the most violent weather.

THE GREAT GRAY-AND-TAN HUNTER IN ACTION

The coyote's many talents all come together when he hits field or forest for the undertaking that occupies most of his waking time: Hunting for food. Now is when he puts his keen senses to full use, and he is fast, agile, smart, and persistent.

Coyotes cover a lot a ground in their nightly quest for sustenance—the average night's journey is about two and a half miles, but it can be as much as ten or more. For the most part, coyotes hunt singly or in pairs. As they walk or trot along, they examine their surroundings closely: watch,

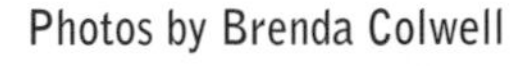

Photos by Brenda Colwell

listen, and turn their sharp noses to the search for anything that might mean dinner.

They have the hunt for mice, which in some parts of the year make up as much as 40% of their diet, down to a science. They listen for the tiny sounds of a mouse moving around down in tall grass or deep snow, or use their sense of smell to target its position. Then, like a cat, they stiffen and pounce on it, or leap into the air and bring their forepaws down hard on the plump little creature. Coyotes can catch a mouse every ten minutes or so this way. And there is no danger of running out, since mice are abundant. In fact, one field mouse, if there were no predators, could produce more than a thousand descendents in a year.

Larger rodents like gophers, prairie dogs, and ground squirrels call for a different approach. Here coyotes often use the "freeze and lunge" technique—waiting patiently somewhere nearby, as long as it takes, for one of these larger rodents to go off guard, or wander closer, and then leaping to snatch it up. They often succeed, but if they miss, they may tumble head over heels in the process. Once a midsize rodent is in their mouth, coyotes often shake their heads vigorously from side to side, to kill the animal and keep it from turning around and putting its sharp "chisel" teeth to work on them.

Coyotes sometimes take on a partner in the pursuit of burrowing rodents. Badgers, who are formidable predators in their own right, are unbelievably fast and powerful diggers. Coyote and badger may go after a ground squirrel or the like in concert—the swift coyote chasing the little animal down into its den, and then the indomitable badger digging down after it. If the rodent decides to use one of his alternate burrows to head back up to the surface, the coyote will be there waiting; if the badger corners it somewhere down below, the prize is his. This odd teamwork goes so far back in time that the hunting buddies, coyote and badger, have been found illustrated on pre-Columbian Native American pottery.

Rabbits, another mainstay of the coyote diet, bring into play other methods. Rabbits, especially jackrabbits, are just about as fast as coyotes, and in some situations, even faster. Coyotes often pair up to hunt rabbits—taking turns chasing them to wear them down, or chasing them up to a wire fence, where they will be stopped or slowed. One or more coyotes may also chase a rabbit up to a third hidden coyote, and coyotes have even been known to employ "distraction action" on prey like rabbits—one member of the pair engaging in distracting antics until the other one closes in. Moving fast, it grabs the rabbit's throat, or its hind leg and then the throat.

In spring and early summer, coyotes put both scent and sight to work in the quest for fawns—watching for signs that a doe has a fawn stashed nearby, and seeking scent trails to spotted youngsters hidden in tall grass or brush. This takes some doing, since does groom their fawns thoroughly to remove as much of their body scent as possible, and fawns instinctively stay as still as possible in their places. Coyotes usually kill fawns by seizing them by the neck, as they do sheep, and when they are done feeding, there may be little left but a few bits of bone and hair.

The badger, the coyote's occasional hunting partner.
Photo by D. J. Hannigan, law enforcement, Park County, Colorado

Larger prey like adult deer, which eastern coyotes, especially, have been hunting more and more in recent years, are a different proposition. A big animal like this calls for a pack approach. To take down a deer, antelope, or elk, three or more coyotes team up. Then they chase it in relays until it tires, chase it up to a fence, or go after it in deep, crusted snow or on ice, where the slim, hard hooves of the deer put it at a disadvantage. Even chases in flat farm fields are often short, 60 yards or less, although coyotes have the patience to keep after a large animal for hours if necessary.

Coyotes often kill deer by slashing the throat, although they may also attack from the rear, sometimes cutting the hamstring tendons in the leg, crippling it. Coyotes have horrified observers by starting to eat large prey, especially, while it is still alive, dragging itself about, bleating and struggling. As a U.S. Fish and Wildlife Service biologist noted once, rather dispassionately, "A predator needs to kill before eating only if the meal is going to move around inconveniently, or . . . cause injury to the predator in the course of its meal." We need to remember here, as noted earlier, that animal predators, coyotes and others, do not glory in their kills, once made. They handle and eat the animal calmly and (as far as human onlookers can determine) dispassionately. Predators have no choice—they must kill to live, and eating a prey animal is about the same as dropping a pork chop into the frying pan, or opening a package of beef jerky.

COYOTE TRACES

Number 1: Tracks

When wondering whether you and your property are being visited by coyotes, or whether you have encountered them in the wilds, tracks are one big clue. Track identification is an important part of animal detective work, although it is always easier in theory than when you are standing outside somewhere, track guide in hand, peering down at a set of marks in the soil or snow.

Here are the main things to remember about coyote tracks.

Coyote Tracks: front foot, back foot. Drawings by Kim A. Cabrera

Shape Consider this first, because it is a key clue. Coyote tracks are oval, whereas most dog tracks are round. They are also more compact—the toes less spread—than the tracks of a dog of comparable size.

Toe count Though coyotes have five toes on their front feet, one of these is a dewclaw, so front or rear, a coyote footprint will show four toes.

Size Eastern coyote tracks are often nearly three inches long, and can be even a little larger than that, and are about two-and-a-half inches wide. Western coyote tracks are two-and-a-quarter to two-and-a-half inches long, and about two inches wide. Eastern or western, the tracks of the hind feet of a coyote are always a little shorter than those of the front feet.

For comparison, wolf tracks are usually about five inches long, and the tracks of large dogs, four or four-and-a-half inches long. Fox tracks are close to the length of the tracks of a western coyote, about two inches in the case of the red fox, but smaller than the tracks of an eastern coyote.

Nail prints Are much finer on a coyote track than on a dog track, and the nails, which point inward, will be most visible, and often quite close together, on the middle two toes.

Heel pads Will be more noticeable on the tracks of a coyote's front feet, and both front and rear heel pads (see illustration on page 37) are a distinctive shape.

Stride The walking stride of a coyote is anywhere from 9 to 17 inches, depending on the size of the animal. The stride of a male is usually larger, as is that of just about all eastern coyotes. Medium-sized dogs, with the exception of greyhounds and whippets, usually have a shorter stride.

The trot stride of a coyote ranges from 15 to 18 inches, or even more, and a running coyote will have sets of tracks that are three to four feet, or even as much as eight feet or more, apart.

Walk pattern Coyote trails are for the most part much straighter than those of the meandering dog, who is usually less aware of his surroundings and usually does not have to "support himself" (get down to the serious business of finding enough to eat) as coyotes do. Coyotes are also famous for "perfect stepping," also called "direct registering," which means that they often put their hind feet down precisely where their front feet fell—a neat trick, and one that requires agility to pull off. If coyotes are traveling in a group, they may all do this, following the path of the first animal, and thus making it very unclear whether you are dealing with one animal or a half dozen.

Some good places to look for them

You will often find coyote tracks along fence lines and trails of all kinds, in draws or gullies, on pond and stream banks, and on ridgetops. Tracks will of course also be found near any carrion that coyotes have been feeding on.

TO GET CLEARER TRACKS You can set up a spot designed for just that purpose. Pick an area where you suspect you have been getting coyote visitors, or one that coyotes tend to favor, such as right by an opening in a fence. Clear a small patch of ground free of vegetation, and level it. Then you can wet the soil down well with water before dark, creating mud, one of the all-time best mediums for clear tracks. Or you can add a thin layer of sand or flour on top of the soil, or sand mixed with a small amount of mineral oil. You can make your track-catching area more attractive by adding some hunter's scent lure, or other bait right by or in it. The U.S. Fish and Wildlife Service uses track plots set up next to scent posts to estimate the population of smaller carnivores in an area.

Number 2: Scat

Coyote scat is something you are more likely to find in a natural area, unless you live in a rural area or have a large suburban property. Coyotes use their droppings to mark their territory, so you are most likely to find them in very visible places, such as on trails, at the intersections of trails, on the top of rocks, large or small, or on fallen logs. Scat may also be found near carrion that coyotes have been feeding on.

Coyote scat is about one inch thick at most, and it tends to be tapered and twisted. And although the color and consistency of it varies with the coyote's diverse diet, it is often dark gray or black, turning lighter as it is exposed to the elements. It is very likely to contain hair, bone chips, seeds, grass, or feathers, and may also have a slightly musky odor. You may find scrape marks in the soil nearby, from where the coyote wiped its feet as it was scat-marking.

Dog scat (which most of us are all too familiar with), on the other hand, is much smoother, softer, and more uniform. Fox scat may be quite similar to the coyote's, except it is often smaller.

Do not handle coyote scat without rubber gloves or a trowel or the like. It often contains the eggs of parasites like canine roundworms and tapeworms. If a human ingests these eggs, and they hatch inside of him or her, the larvae that come forth may end up in parts of the body they would never be in, in their normal host. This can have serious or even fatal results.

Number 3: Coyote sleeping beds

Coyote sleeping beds are most likely to be found in the winter, when, like the places where deer have bedded down, they are made obvious by the fact that snow did not fall in that spot. In winter coyotes curl up in a circle, with their heads covering their feet and their tail over their nose, and these circular beds, up to two feet in diameter, will often be found in high spots or places with good cover. Coyote beds, compressed circles in grass or other vegetation, will be more oval in warmer parts of the year.

CHAPTER 3

a look at coyote society

This chapter will take a quick look at the social habits of coyotes and will answer such questions as: Are coyotes solo or pack animals? How much land constitutes a coyote's home territory? What does a typical coyote family consist of? When do they breed and how do they raise their young? Do they actually crossbreed with dogs? How might you find a coyote den? Addressed here also are the myths and truths about coyote serenades—why they happen and what they mean.

DO COYOTES ACTUALLY FORM PACKS? WE OFTEN SEE JUST ONE, OR MAYBE TWO OR THREE TOGETHER

There is some difference of opinion on whether coyotes form packs, partly because, as naturalist Hope Ryden once pointed out, "In the area of social relations, more has been learned about the rare wolf than the relatively prevalent coyote." The coyote's furtive and elusive nature makes observation of coyote community life difficult, unless you're willing to spend weeks and months hiding out near coyote hangouts, as Hope was.

Though some scientists feel that coyotes have a looser and less formal pack structure than wolves, as more research has been done in this area, coyotes look more and more like true pack animals. One of the big reasons for wolf packs is that wolves mostly feed on large animals that take cooperative effort to bring down. Coyotes often feed on smaller animals that they can catch alone, but even so operating as a group, at least at times, has some real advantages for them. The size of a coyote pack can vary in different parts of the country, and in different types of habitat. But coyote packs are usually family groups, composed of about six or so animals—a mated male and female, plus their pups of the year, and often a couple of last year's young, called beta, helper, or associate coyotes, who stay around and help the parents feed and raise the pups, and help defend the pack's territory. The members of a pack can recognize each other, even at some distance. They aid and defend each other, howl together, and engage in a very precise series of rituals, as wolves do, that acknowledge and preserve each animal's status in the group. These rituals also help prevent

Dominance is an important issue in the coyote hierarchy. The dominant female (on the left) intimidates a young male (on the right). Photo by Glenn D. Chambers

serious disputes between members—and invading coyotes from other packs—from breaking out.

Aggression and dominance are communicated by such things as raised hackles on the back and elsewhere on the body, head held high with neck arched, ears pointing forward, tail fluffed out and held at about a 45-degree angle, and most impressive of all, mouth opened extra wide into what is called a "gape," giving the coyote a very nasty look and exposing a formidable set of canines and incisors.

Submission, on the other hand, is signaled by such actions as retreating, rolling over to expose the stomach, urinating, whining, tucking the tail between the legs, flattening the ears, and coming forth with a slightly different version of the gape.

The highlighted coloring of coyote fur patterns, such as the white hair by the ears, the white borders around the eyes and lips, and the long black guard hairs in a coyote coat, help dramatize the gestures that broadcast status and intent.

Not all coyotes are members of packs. A goodly number of them are transients, solitaries, or floaters—coyotes who are wandering about looking for a territory of their own. These solitaries travel large distances and do not breed until they find a niche for themselves.

Were coyotes more gregarious earlier?

Accounts from the early days of the exploration of the Midwest and West include descriptions that make it sound as if coyotes assembled in greater numbers back then. The huge amounts of carrion made available by white buffalo hunters and the like may well have attracted scores, or even hundreds, of coyotes and other predators. It has also been theorized that as white settlers took over the country, coyotes found it prudent to become less social (and thus less noticeable).

Even today, large numbers of coyotes sometimes turn up, especially in winter, to take advantage of exceptional food sources, such as a concentration of large dead animals. These are only temporary gatherings of coyotes from different territories, which will break up when the food is

Coyote showing the fearsome coyote "gape," designed to warn off other coyotes. Photo by Steve and Dave Maslowski

gone. In some areas a bunch of teenage coyotes may stay together for a while, and in the West, especially, groups of male coyotes may travel together behind a female approaching heat. But these too are only short-term groupings.

In fall, most family-based packs will undergo some changes as some of the new pups, and perhaps some of the beta coyotes, head off to find homes of their own. Among eastern coyotes, which hunt a lot of adult deer in winter, there is a tendency for family groups to stay together longer, to help with the hunt. Some researchers also feel that eastern coyotes are more social in general, thanks again, perhaps, to that infusion of wolf genes.

THE COYOTES HAVE BEEN MULTIPLYING . . . SOON WE'LL BE OVERRUN WITH THEM!

There are a number of reasons why it can seem as if the coyotes in an area have increased. For one thing, coyotes exist in many more places now than they did decades ago, so the very sighting of them for the first time by many people, which is a novelty, makes them stand out. Then too, when young coyotes leave their parents in the fall, and start traveling around looking for their adult homes—they can journey as much as hundreds of miles in this quest—it may seem as if there are more coyotes around. The fact that coyotes everywhere travel up to ten miles in a day or night, hunting for food and perhaps howling, can also give the impression that the woods and fields are filled with coyotes, when it may just be the same small band, moving around.

Coyotes actually have their own built-in system for preventing overpopulation: they establish and maintain well-delineated territories, and defend them against all other coyotes. The pups of a mated pair will share this territory when they are young; it will be their playground and training ground. In some parts of the country, such as the upper Northeast, the pups may even stay on in their birth territory for a year, helping their parents and younger siblings catch deer. But eventually they will have to strike out to find a territory of their own. Rarely does an extended family of more than five or six adults plus their pups share a territory. A coyote's territory is where he sleeps and breeds and does his hunting, and he knows every trail, roadway, landmark, and patch of cover in it.

How large are coyote territories? Anywhere from the smallest noted to date, one quarter of a square mile, in the city of Los Angeles, to 30 or even more square miles. If that sounds like a lot, consider the fact that wolves have a home range of 100 to 200 square miles! Researchers do not fully agree on this, but in general urban and suburban coyotes have smaller territories than rural ones, and the density of coyotes (total number of them per square mile) is heavier in urban environments. Coyote territories in the city of Chicago, for example, are estimated to be from 5 to 10 square miles; in urban parts of Massachusetts, they average 11 square miles. Many sources say that the territories of male coyotes are larger than those of females, but female territories are more exclusive.

The amount of food available has a lot to do with the size of a territory, which is one reason coyotes are able to support themselves within a smaller section of goodies-rich urban and suburban environments. Coyotes may need a range of 15 square miles in arid rural country, and only five or so in parts of the same state with rich woodland and lots of rabbits and livestock. In winter coyote territories may be expanded to twice their size, since food is harder to come by; for this same reason southern areas with no harsh winters can support more coyotes per square mile year-round. Eastern coyotes generally have larger territories than western ones, since western coyotes have a larger array of small mammals to feed on.

The coyote core

Within every coyote territory is a much smaller core area that serves as the main base of operations. This carefully chosen headquarters, which may be 10% or less of the whole territory, is where the resident coyote or coyotes spend most of their time. Which means it must have a good supply of food and water, good places to bed down and build dens, a good view of the surrounding terrain, and some protection from predators.

Signals to stay out!

Coyote territories have clearly established boundaries, some of which may be very clear-cut features such as fences, roads, bodies of water, or trails.

Coyotes do not normally defend their territories with fang and claw, to the point of death, as wolves do, but they do work hard to keep unauthorized members of their own species out. This is accomplished in a number of ways, but scent-marking is the big one. Any owner of a male dog or tomcat has a good idea of what this is all about. The coyote owner of a territory, usually the dominant male, makes regular rounds of prominent places within, and on the boundary of, his territory, and makes his mark. This means lifting his leg to spray the spot with urine, or in the case of females, squatting to do the same. The objects anointed for this purpose may be stumps, rocks, fence posts, or bushes—almost anything that stands out in the landscape. Coyotes use their droppings similarly, though these are just as likely to be on level ground. They are able to secrete a strong-smelling pasty substance onto their droppings from glands near their anus, to make the message unmistakable. This anal discharge will tell exactly what coyote left his mark, and why.

Howling is another way that individuals and groups advertise their possession of a territory, and it, like scent-marking, has the advantage that trespassing coyotes can be warned off without ever confronting an indignant resident face-to-face. This helps prevent bloodshed, even though coyote patrollers, as noted earlier, are more likely to snap at, make nasty faces at, and chase intruders than to sink their teeth into them.

I'm going to get rid of every one of them!

The fact that coyotes, as noted above, carve out and occupy very specific territories, is one of the big reasons that you can't "wipe out" all of the coyotes in an area, or on your farm. The fields and woods (and now the streets of nearly every city) are full of "transient" coyotes that have no territory of their own yet, though they are on the move ceaselessly in search of one. The minute a territory holder dies, or is exterminated, one of the transients will rush in to fill the gap. It has been estimated that vacant territories are filled by traveling loners within days, or at most a week, of the time that a subtle "room available" sign is put out on the coyote grapevine. And most of these are younger animals, which means a higher reproductive rate that increases the overall population.

THEIR CLAIM TO FAME: COYOTE HOWLS AND OTHER SOUNDS

Very few wild animals are heard much more often than they are seen, but the coyote is one of them—one of the very few in North America. Far more people have heard coyotes, whether they realized it or not, than have seen the authors of those exotic sounds. Wolves also howl, but only a tiny percentage of the population is in a position to hear them, except in scary movies.

The coyote's vocal talents are what gave it its name—*Canis latrans,* or barking dog (although one wonders why it was not "howling dog," since the coyote's howl is far more distinctive and unmistakable than its bark). Its nightly serenades have also caused it to be dubbed the "prairie tenor," and are why Native Americans call it the "song dog."

The coyote makes close to a dozen different sounds, but is best known for its howling. Many people have tried to describe it in words, and the description that comes closest, for me, is a high-pitched, long, wavering yodel or tremolo. The eastern coyote is said to have a deeper voice than his cousins elsewhere.

Often coyotes howl as a group, and when this happens, it is done in harmony, each "singing" on a slightly different high-pitched note. The howls blend and swirl together, preceded, and often followed, by a series of short yips and yaps. When you hear an eerie chorus like this, or even the lone howl of a coyote, the hair on the back of your neck may stand up, and you pay attention. Some have speculated that this is an instinctive reaction dating back to the days when we humans were prey far more

Photo by Steve and Dave Maslowski

often than we are now, or as Charles Cadiuex, author of *Coyotes: Predators & Survivors,* put it, "an atavistic reminder of the days when our ancestors cowered in caves, listening to the sounds of the hunting packs."

The coyote's howl has been honored with such titles as "the voice of the West" (now an outdated reference) and "the call of the wild." It has been poetically summarized and complimented often. Lila Lothberg called it "the saddest, most beautiful, and most triumphant music in nature," and J. Frank Dobie said that the coyote's howl "speaks of old, unhappy, and far-off things, and of the elemental tragedy of life. . . . Its dark . . . beauty has taken me far away and filled me with a sense of the mysteries." The coyote howl has such a mystical and thought-provoking effect on people that it has been theorized that perhaps state and national parks should arrange for a way for visitors to experience it. Some in fact do, with evening nature walks led by naturalists.

They're right over there . . . or there

Adding to the coyote's mystique is the fact that it is devilishly difficult to determine exactly where coyote howls are coming from. At first you are sure they must be right behind the barn, but when you race out there with your lantern, cast your light in every direction, and fail to see any trace of them, you sit in the dark and wait, only to realize that no, they must be over there to the east. Or maybe actually closer to that fence line at the end of the valley. . . .

Even seasoned coyote trackers and hunters, after listening to a set of coyote sounds and then pooling their impressions, have all ended up pointing in different directions. This phenomenon has added "ventriloquist" to the list of coyote talents, and it is at least partially caused by the following: as the sound travels, especially in hilly country, it bounces off ridges, boulders, rock outcrops, and lines of trees and buildings, so that what you hear is partly the original sounds and partly echoes. This, plus the varying tones of the howls, and the fact that some animals switch frequencies in the middle of the songs, also makes it hard to tell just how many animals you are hearing. In general, the answer is fewer than you think. What may sound like a giant pack could be just a pair plus some pups, or maybe the pair alone. Scientists have speculated that this auditory illusion evolved as a way of making a group of coyotes sound more formidable to other competing groups of coyotes.

When are they heard the most?

Coyotes are heard most often in mating season, January and February especially, in fall when the pups strike off on their own, and on cool, clear nights in early winter. They howl the least in the early stages of pup-raising. Coyotes are also often heard right before, or right after, the passing of a weather front.

Dusk is the time of day you will most commonly hear coyote choruses, followed by various times of night, and daybreak.

Some coyote groups, and individuals, howl much more than others.

How they go about it

Coyotes often seek out a spot, such as an elevated and/or open area, from which the sound will carry, and howls can often be heard for three miles or more. They raise their heads when they howl, and control the pitch by varying the size of their mouth opening. When coyotes do a group howl, they all stand and wag their tails vigorously.

Parent coyotes tutor their young in howling, and pups only a few weeks old have been known to howl.

Why do coyotes howl?

For a number of different reasons, and many of these are the same reasons wolves howl.

- During mating season, to attract or seek mates. It is even speculated that the amount of howling the coyotes in an area do around this time, and right after it, can affect the size of the eventual litters, by causing some fetuses to be reabsorbed (see page 54 regarding the way in which coyotes can vary their reproductive output).

- To assert their claim to their territory, and warn other coyotes to stay away. This has the advantage of preventing direct confrontation with competing groups or individuals, and the fights and scuffles that might result.
- To promote pack harmony and solidarity. Coyotes will often do a group howl (those howls at dusk) before they start in on their nightly hunting. Group howls are led by the dominant male, and it is thought that group howls somehow also reaffirm the status of individual coyotes within their group.
- To locate each other when pack members are separated, or herald the fact that they have reconvened.
- To celebrate when the pack has made a kill or to let other pack members know that easy prey has been found.
- When coyotes have pups in a den, the adults may scatter, if an intruder approaches, and howl, to draw attention away from the den.
- Coyotes, like wolves, can recognize the voices of individual pack members. And like other animals, coyotes probably use nuances of the noises they make to convey information to one another, which we cannot decode since we have not yet found the coyote Rosetta stone.
- Even some scientists believe that coyotes howl at times just for the fun of it, to express their pleasure in just plain being alive.

Other coyote sounds

Coyotes don't just howl, they also bark and growl, usually as warnings to one another, or when surprised. They make a kind of low whine as well, used to greet pack members or to demonstrate submission to an animal higher on the pack social ladder. A sound called a "huff" is made by expelling air from their mouths and noses, and is often used to summon or warn pups of danger quietly. Coyotes sometimes yip when pursuing prey and make an assortment of other odd noises that include "laughs" and "gargles," which are reminiscent of the sounds made by hyenas. These sounds, plus their howling, may be why the famous naturalist Ernest Thompson Seton called coyote songs "a Wagnerian opera of diabolic strains."

Sounds that will provoke howling

You don't have to be a coyote to start a round of howling. Coyotes have been known to start howling when they hear fire or tornado sirens, train whistles, low-flying aircraft, ambulance sirens, and other high-frequency sounds. Hunters sometimes take advantage of this to scout out coyote daytime bedding places before a hunt, by going around with a small siren or coyote howling call to see where they get answering howls. People who have pet coyotes, or who simply like coyotes, have been known to howl to coyotes, tame or wild, and be joined or answered. And the coyotes have always been polite enough not to criticize their technique.

HOW COYOTES KEEP MULTIPLYING: ONE + ONE EQUALS HALF A DOZEN... OR MORE

After more than a century of the most energetic extermination campaign ever waged against a species on this continent, coyotes are not endangered, not threatened, not even reduced. In fact, as noted earlier, they are only multiplying. One of the reasons for this is their skill in the bottom-line survival undertaking called reproduction. Coyotes are exceptional parents, and they have a number of other instincts and habits that serve them well, upping their score in this numbers game.

Among the first of these is the strength of the male–female pair bond, which is the cornerstone of coyote society. No, they don't mate for life, but they do come closer to that ideal of romance fiction than most of the other animals often cited as an example of this, such as wolves and foxes—or for that matter, humans. Coyotes choose their mates carefully and stick with them throughout the entire breeding process, and often several breeding seasons thereafter. A mating between Ms. and Mr. Coyote is no one-night (or one-hour) stand, but a partnership in which both will work hard to ensure the survival and success of the fruits of that union—the pups that the mother will eventually bring forth. Coyotes have been known to start this mate selection process when they are just "subteens" of six months or so in age.

Another big reason the coyote reproductive process is so successful in keeping coyotes thriving is that it has a built-in expansion and contraction factor. When the coyote population in an area is already heavy, smaller litters are produced. When food is plentiful or coyotes are being heavily hunted or otherwise eliminated, litters get larger.

Heat in a cold time: The coyote breeding season

Female coyotes are sexually mature at the age of ten months, but they don't usually breed until their second year. Once it starts, coyote romance occurs within a tight time frame: the females are only sexually active about two months out of the year (the actual heat period is only a single week), and the males only produce sperm for at most a third of the year.

The whole process starts in January or February (earlier in southern areas), a cold time in most parts of the country. As the female goes into the first stages of heat, her vulva swells and bleeds, and she begins to attract a string of would-be mates. Snow, sleet, or chilly rain notwithstanding, her admirers follow her from place to place, and the night is lively with howls—plus a few snarls and growls. But the suitors are peaceful enough otherwise, and they may skimp on little details like eating and sleeping in their eagerness to stay alert enough to be "the one."

Finally, she makes the choice, which is often based as much on compatibility and unknown "X" factors—this is a very personal, individual decision on the female's part—as on how much he weighs, how tall he is, or how lustrous his coat is. With a variety of playful gestures, including the licking of his chin and nudging of his chest, she gives him the good news.

They then spend a few days cementing their relationship—traveling together, sleeping by each other, and howling in harmony.

The clock is ticking now, however, because that single week of actual heat has arrived—the only time she will allow a mating, and she ovulates during the last two or three days of this week. So the female's reluctance and evasion is finally exchanged for direct prodding and hinting on her part, if needed, such as standing in front of him and moving her tail aside. Once the male mounts her, he thrusts his penis deep into her vulva, until its swollen base is gripped hard by her strong inner muscles. This is followed by the famous copulatory tie of dogs and their relatives, when the male and female, still connected, stand tail to tail for 20 to 30 minutes afterward, as his sperm are slowly transferred to her—a highly vulnerable position to be in, in field or forest, to say the least. The pair copulate a number of times in the days that follow, until the bared fangs of the female indicate that heat is over for another year.

Nursery hunting

As the female proceeds through her pregnancy, which lasts for 60 to 63 days in western coyotes, and a little longer, perhaps as much as 65 or 66 days in eastern coyotes, the prospective parents start looking for a nursery site. Though coyotes never live in burrows or holes of any kind, they do use them to shelter the pups for the highly vulnerable first six weeks or so of their young lives. A good den site usually faces south, so that the mother and her youngsters will get the benefit of spring sunshine, and it is located within a half mile or less of a dependable water source. Ideally, it is elevated to some extent too, to protect it from flooding and to allow for reconnaissance of the surrounding territory. It should also have brush or woodland nearby to provide cover when needed and, above all, to be as far away from human disturbance as possible.

This being said, coyote dens are found in an amazing variety of places. Among the most popular are sandy hillsides, knolls, and stream banks, and just beneath rocky outcrops or ledges. Often chosen, too, are log piles and brush piles, fencerows, thickets, trees hollow at the base, and spots under

A female in the den with her babies. At one week of age, these coyote pups nurse at least three times a day. The female attends them almost constantly at this early age. Photo by Glenn D. Chambers

fallen logs or dense bushes like multiflora rose. Dens are also found in more urbanized settings, such as under sheds or granaries, in culverts, even under the stones and crypts of cemeteries.

The coyote pair picks not one but up to a half-dozen den sites, and readies them all ahead of time—and this is a lot of digging. They may make the job easier sometimes by simply enlarging the former homesite of a fox, skunk, groundhog, badger, or even porcupine. There are a variety of construction designs, but they all usually include a long burrow from the opening that goes down to a nursing chamber about three feet wide. The den almost always has a second entrance, one that is much better concealed than the main entrance. If you come across a den in the woods and wonder if it is a coyote's, be aware that the opening of a coyote den is normally from 13 to 24 inches wide.

The moment of truth

First-time mothers sometimes give birth as late as June, but for the most part, coyote mothers deliver between mid-March and mid-April. Depending on when she conceived, the little over two months of pregnancy will be over, and the female will go into labor down in the darkness of the den. After three to ten hours of grunts and contractions and the steady emergence of slippery little bodies, the litter will be born. Once in while a mother coyote comes forth with a dozen or even more babies, but the average is five or six.

The pups, little bundles of short, soft gray or brown fur, are blind and helpless at birth, like dog pups or kittens. They can't even regulate their body temperature when they first emerge. Within two weeks, their eyes are open and their ears stand up, and within a week after that they are walking, and then running. When the pups are three or four weeks old, they crawl up to the big world outside the den, under the watchful eyes of their parents, and by the age of six weeks they are sleeping above ground, except during bad weather.

A few minutes after birth the pups are nursing enthusiastically, and this continues until their emerging sharp little teeth start nipping the mother, and they are weaned at the age of five to seven weeks. When they are only two or three weeks old, their mother starts supplementing their milk diet with semisolid, partially digested food that she eats and then regurgitates for them. When they are a little older, the whole coyote family, mother, father, and any "helper" or beta coyotes, will start bringing them small animals like mice and rabbits, and pieces of meat and bones.

Pup perils

Regardless of the size of the litter and all the care their parents lavish on them, less than half of coyote pups, and often only about a third of them, survive to see their first birthday. Like all baby animals, they are at risk from predators, in this case eagles, large hawks and owls, bobcats, lynxes, wolves, mountain lions, bears, badgers, wolverines, large dogs,

Coyote pups are way up there on the cuteness scale. These two ten-week-old pups still nurse occasionally but feed primarily now on food brought to them by their parents. This includes carrion as well as freshly killed rodents, small birds, or rabbits. Photo by Glenn D. Chambers

and rattlesnakes, plus human hunters and trappers. Adult coyotes from other packs may kill pups, too, if they happen upon them, or the pups are caught trespassing.

Diseases and parasites—including distemper, canine hepatitis, parvovirus, mange, roundworm, and anemia from hookworm and tick infestations—kill many pups as well.

Den switching

At the slightest sign of human disturbance, predator menace, or danger of any other kind, the parents will grab their pups and transport them to one of the alternate dens they so thoughtfully prepared months ago. Coyotes have been known to move their dens a half-dozen times, and sometimes they may do so just to leave behind the fleas and feeding fallout a den site can accumulate. Coyotes are very sensitive to intrusion upon their pup-raising—captive coyotes may kill the entire litter if people interfere with it or approach it too closely in the early stages of parenthood.

Pup training camp

By midsummer, the parents often move their brood to a different kind of nursery site, this one a kind of halfway house or training camp for independence. This is frequently an elevated area, too, with good visibility in all directions, plus water, cover, and shade for the pups. This is called a "rendezvous site" by coyote and wolf experts, where the pups will be left while the parents hunt, and from here the youngsters may even launch their own little hunting expeditions to nearby places. Parents by this time have given and are giving many lessons in the art of keeping yourself fed.

Coyote pups, like domestic dog puppies and kittens and the like, have a whole repertoire of games they play with each other, everything from tag to jump on and wrestle to nip and flee. Two things are distinctively different in the case of young coyotes. They are much more aggressive at an early age than wolves are, and by the age of four or five weeks, are having not-so-mock battles with each other to establish dominance. Coyote pups also signal each other before they start to play, making it unmistakably clear that this is a play encounter.

Leaving the nest

By fall, when they are six to nine months old, the pups are getting ready to leave the nest, to go out and seek their very own territories, and eventually, mates. By now they have learned to hunt, and have made longer and longer excursions of their own in preparation for the big day. A few of the pups may, and often do, stay with the parents a while longer, to help them hunt in winter and raise next year's litter. The experts seem divided as to whether those that stay are always females that bond well with their parents, or the

Two coyote siblings, about six months old, in Camas National Wildlife Refuge, Idaho. Photo by John H. Williams

most dominant of the pups, as determined by the dominance competitions of puphood.

A few adolescent coyotes may stay in their birth territory, too, without closely associating with their parents, and these have been called "slouchers," in view of the fact that they resemble grown human children hanging around at home without lifting a finger to help, or paying board either!

But by the time the leaves are falling, the majority of the litter will be setting off to make their way in the world, traveling 20, 50, or even 100 miles away to keep those coyote colonizations ever advancing. Here, too, the survival rate is low, 50% or less, since these are young, inexperienced animals on the move, and possibly facing situations and dangers they never did at home.

WHAT ABOUT COYDOGS (CROSSES BETWEEN COYOTES AND DOGS)? IS THERE REALLY SUCH A THING?

There is, but they are not nearly as common as many people seem to believe. There is not a coyote hiding behind every fence and shrub, waiting to pounce lustfully upon your poodle or Pomeranian. (One person interviewed for this book even believes that coyotes intentionally seek out hunting dogs in heat, for the express purpose of adding more useful genes to their heritage. Coyotes are smart, but probably not into conscious genetic engineering just yet.)

Coyotes want to mate with coyotes, of which there is usually no shortage. Though they do mate with dogs at times, it generally only happens under unusual circumstances, such as when coyotes are expanding into a new area, and the chance of finding a fellow—or female—coyote is slim.

There are other reasons for which coyote–dog crosses are not happening constantly. Dogs and coyotes are not necessarily friendly to one another on sight—they are more likely to fight, chase, or be afraid of one another, than to engage in doggy horseplay . . . or foreplay. Coyotes made captive for crossbreeding research purposes have shown a real reluctance to mate with dogs.

Then, too, coyotes have sex on their minds for only a small part of the year—about two months, somewhere between late December and

March. Males come into their yearly three months or so of fertility during this time, and females into their single yearly heat. Female dogs, on the other hand, come into heat twice a year, and male dogs, like human males, are always able. But this means that windows of opportunity for dog–coyote crosses are limited.

Coyotes are pickier than most dogs, as well, when it comes to mate selection. The females, especially, give the matter a fair amount of study, and once a suitor is selected, the pair stays together during the female's pregnancy and the birth and rearing of the pups, and often for years afterward. Not only would a dog have to make the cut here, but he or she would also have to give the right courtship signals and messages—not necessarily the same in dogs and coyotes—during the courtship and "getting-to-know-you" interactions. The dog would also have to be large enough to avoid being seen as dinner (see Chapter 5).

If the odds above are overcome, and the unlikely does occur—a female coyote crosses with a male dog to produce coydogs, or a male coyote with a female dog to produce "dogotes"—these offspring will have their share of problems. A female coyote pregnant with coydogs will have no mate to share the care of her young. Dogs, oddly enough, do not demonstrate what is thought of as perhaps the most key features of their wild relatives: a strong pair bond between mated male and female, and the male's willingness to help raise the family. And once a litter of coydogs is born to their "single-parent" mother, they neither will fit well into their mother's wild lifestyle, nor make good pets for people. Coydogs are often larger than coyotes, and less afraid of humans, and they often have some of dogs' less admirable habits, such as running and chasing other animals for fun, which makes them a greater danger to livestock and game animals.

True, the Internet does contain some touching testimonials to the friendliness and cuddliness of a few individual coydogs, but for the most part pet coydogs are shy and fearful, and can be quick to bite when they feel afraid or threatened. The owner of a pet coydog needs to bone up on, and master, being a pack leader and alpha dog—firm, confident, and utterly consistent authority is essential. If a person gets a coydog when it is young and trains it carefully, the animal will do better—but one must wonder: except for the ever-present lure of exoticism, why go to all this trouble, and perhaps put yourself in danger, when scores of breeds that make much better pets are available?

Coydogs, though a hybrid animal like the mule, are fertile. But they come into heat in the fall, and have their litters in the winter, which in most parts of the country means severe weather and scarce food. This is not a favorable situation, especially since male coydogs, like male dogs, do not lend a paw with pup-raising. This breeding timetable, however, does keep coydogs from being able to mate with coyotes and create an ever more doglike mixed race of wild canids on our doorstep.

To be absolutely certain whether an animal is actually a coydog, its skull must be examined by scientists who know the difference between dog, wolf, and coydog features, or better yet, its DNA tested. What do coydogs look like? Like mongrel dogs, they will usually look something like each of their parents. But larger than usual upright ears (like a coyote) and "very piercing eyes" are often reported. And some coydogs look very much like coyotes.

Helping to further reduce the population of coydogs is one more reason to spay or neuter your dog.

Photo by Steve and Dave Maslowski

CHAPTER 4
the country coyote

This chapter looks at coyotes in rural, wildland, and farm areas from a rather specific perspective: the kinds of mischief they can get into there and the types of problems they are often accused of. This includes their effect on other, smaller wildlife, on deer and pronghorn, and the big, hot topic: to what extent they prey on sheep, cattle, and other livestock. You will find here also a guide to predation diagnosis, to help you determine whether coyotes are responsible for the damage in question, or some other wild animal or free-running pet is to blame.

WHAT EFFECT DO COYOTES HAVE ON OTHER WILDLIFE?

That subject in its entirety could fill a bookshelf, but let's look at some of the species people worry most about, and/or upon which coyotes have had the most obvious effect. Deer and other large hoofed wild animals, which are often thought of first when this question arises, we will look at later in this chapter.

Wild turkeys and other ground-nesting birds

Wild turkeys are a big concern. Even nonhunters are happy to have this handsome and very capable big bird restored to the woods and fields of many states across the country today, and no one wants to see their numbers diminished. More wild turkeys succumb to predators than anything else, at every stage of their lives, with the exception of adult gobblers, whose greatest enemy is human hunters. And many hunters have reported coyotes suddenly materializing the minute they start turkey-calling.

Coyotes do prey on wild turkeys—from turkey eggs, to the hens sitting motionless on them trying to hatch a brood, to turkey chicks (called poults), to gobblers—but not to any greater extent than other predators that enjoy a turkey dinner. This includes raccoons (the most notable attacker of turkeys), bobcats, dogs, owls, hawks, and rat snakes. Wild turkeys, the adults of which are not easy prey for anything, are not a major menu item for coyotes, and when coyotes are removed from an area, losses to other predators simply increase. Only turkey populations that are severely stressed by other forces, such as loss of habitat, disease, or poaching, are likely to be adversely affected by coyote predation.

Coyotes also prey on quail, grouse, ducks, and other ground-nesting birds, but here, too, no worse than other predators. In fact, in some situations, they actually benefit such birds, by reducing the number of smaller predators that specialize in seeking out nests and eggs. And in the case of sage grouse, by preying on rabbits, which compete with the grouse for many food plants, coyotes aid the grouse population. When it comes to these other types of ground-nesting birds as well, removing coyotes from an area often increases losses to other predators.

Other smaller predators

Speaking of those other predators, the fox is foremost among them. Although as Hope Ryden noted, "in behavior as well as appearance, coyotes often seem more foxlike than wolflike" (she was referring, we may note, to the western coyote, although this is true to a lesser extent of the eastern), coyotes are in fact mortal enemies of foxes. The handsome and storied red fox, and the gray fox, too, have become far less numerous wherever coyotes have moved in and multiplied. The reason for this, as the scientists explain it, is that the top predators in an ecosystem usually harass or kill the "mesopredators." Thus where wolves and cougars exist, they kill coyotes, and where the coyote is the biggest predator on the block, he chases or kills his competition, such as foxes. This is easy for him to do, since a coyote is about three times the size of a fox. Coyotes will eat young foxes, and are likely to kill any fox they can corner, or find in a trap. Foxes do manage to survive and continue to exist in coyote country by staying clear of coyote travel lanes and core coyote territory. They are a little safer in the suburbs, where they have more places to hide.

As for bobcats, although a male bobcat would be a formidable opponent for a coyote, and bobcats can climb trees, coyotes have been known to kill young and female bobcats, and occasionally to hassle a bobcat they encounter. They will also kill bobcats caught in traps. But the fact that the bobcat population is usually reduced when coyotes become numerous is mainly because both of these animals feed on much of the same prey.

Varmints

Those not always fully appreciated critters known as raccoons, opossums, skunks, and groundhogs—coyotes eat all of them, especially when they are young, tender, and unwary. But we still have more of them in our backyards, barnyards, trash-can enclosures, and chicken yards than ever. Except for the groundhog. Groundhogs do not have a tremendous number of fans either, especially among farmers and gardeners, but there are notably fewer big, plump, brown rodents seen standing and munching in our fields and on our roadsides since the great coyote invasion.

Threatened and endangered species

In some situations and areas, such as islands (always a very vulnerable habitat), coyotes do pose a threat to threatened or endangered species, in the sense that they are one more stress, in the form of an ultra-efficient predator, added to an already precarious environment. Thus coyotes have been removed or reduced in places where they are a danger to birds and animals like the piping plover, least tern, burrowing owl, and kit fox, and sea turtle nests.

Feral dogs and cats

Not everyone would consider these animals wildlife, but they exist in considerable numbers across the country, and they have a big impact on other wildlife species. In the areas to which coyotes have spread in recent years, there has been a definite reduction in feral cats and dogs. Many people see this as a plus, since, for example, more than one third of the diet of feral cats is songbirds.

Ravens

Ravens, or those big black birds larger than crows, made famous by Edgar Allen Poe, have increased since coyotes headed east, as they now again have a big predator likely to supply them with leftovers. Their former benefactor was the wolf, now an exotic species confined to a few parts of the West, Canada, and Alaska.

What about deer and pronghorn? Are coyotes reducing their numbers?

(By "deer" here I mean whitetail deer and mule deer.) This is a highly charged question, and the people most keenly interested in the answer tend of course to be deer hunters. We humans, the most effective and dangerous predator of all, can be very touchy about the idea of other predators horning in on what we consider our rightful harvest of wildlife.

The first part of the answer is that coyotes are unquestionably killing deer, and more than they ever did in the past. Killing an animal as large as a deer, that can run fast and has sharp hooves and perhaps antlers as well, is not a one-coyote undertaking. It usually takes a pair, or ideally three or more coyotes to accomplish this. Coyotes have always fed on deer that died of natural causes, were killed by vehicles, were only wounded by human hunters, and on gut piles left behind by hunters field-dressing game. But in recent years, coyotes have increasingly turned to finding their own fresh venison.

Eastern coyotes in particular, with their larger body size, larger jaws, and other input from wolf genes, have taken to hunting deer in small packs. Family groups are staying together longer into the year to accomplish this. As far back as 1956, deer were determined to be the primary winter food of coyotes in Michigan, and studies in the Adirondack Mountains of New York have found deer to compose 80% or more of coyotes' winter diet. Anywhere that deer are likely to encounter heavy snow in winter that hinders their movement, they are very vulnerable to coyotes. Coyotes will even take down an occasional elk or moose.

Deep snow gives coyotes an edge in deer hunting. Photo by Colorado Division of Wildlife (Leigh Gillette)

Coyotes do prey on adult deer at other times of the year too, but not as much as in winter. And evidence shows that they are not, in the old time-honored view of the predator, taking only the lame and aged, but equally, healthy deer in their prime.

The great fawn search

Mother deer are fierce protectors of their young, but they do have to leave their little ones alone—hidden in the grass or other vegetation—at

Deer fawns are a large part of the coyote diet in spring and early summer. Photo by John H. Williams

Pronghorn fawns are also preyed upon by coyotes. Photo by John H. Williams

times so that they can graze themselves. In spring and on into summer, the seeking out of unattended newly born or very young fawns has become a coyote specialty. In many areas of the country the fawns lost in this way can be as much as 75% or even more of the annual crop of "Bambis." This does indeed slow the growth of the deer population, and in some areas, it can actually reduce it.

Researchers have actually watched pronghorn females hide their young. As soon as the female walks off a ways, coyotes approach and devour the youngster. The game management arms of different state governments have conducted experiments that have proven that if predators, particularly coyotes, are removed or reduced in an area, fawn survival increases dramatically. (Such studies have also noted that when habitat is poor, predation rates increase.)

Does all of this really harm the deer herds?

For the most part, the answer to this bottom-line question is "not really," although coyote depredations may have a serious effect in a few areas where deer are really struggling to get by, because of other stresses.

Deer today have more than predators to contend with. More whitetail deer, for example, are killed in collisions with cars than by coyotes. Whether or not there is enough available food for them in an area—the quality of the habitat—also affects them more than coyote predation. Human policies and activities, such as the endless transforming of rural land into yet more housing developments and shopping centers; deterioration of habitat by grazing sheep, goats, and other domestic stock; and forest management policies that prevent the wildfires that improve woodland forage for deer, are all detrimental to deer, as is the chasing of deer by feral and free-running domestic dogs.

The biggest threat of all to whitetail deer today is their own numbers. To fully appreciate this we have to take a quick look at Whitetail History 101. Before European settlers arrived in this country, there may have been as many as 40 million or more deer here (coexisting successfully, we may note, with not only coyotes but other large predators such as wolves and cougars). By the beginning of the 20th century, unrestrained killing of deer for their hides, as well as market hunting, had reduced the national whitetail herd to a mere half million or so. Then market hunting was outlawed, and great numbers of farms were abandoned during the Depression. Those farm fields grew up into brush and saplings, and the whitetail, an "edge" lover like the coyote, again flourished, with the help of hunting regulations designed to aid the cause. Estimates vary, but today, since the advent of wildlife management, there may be as many as 25 to 30 million deer jumping fences and waving their lively white tails across the country. But there are less wildlands today to support and contain those numbers.

So in many areas, there are actually too many deer, and this causes a number of problems. Deer feed mainly on young trees and shrubs, and

A coyote at Yellowstone checks out an unattended baby bison. Photo by Bob Melin

just like us, they eat the stuff they like best first. When the deer population is high, over time this reduces the diversity of a forest, as some species in an area may be consumed out of existence. Deer eat understory vegetation, too, which reduces wildflowers and other desirable plants, and their relentless munching of trees and shrubs can destroy reforestation attempts and even affect songbird populations. Deer in suburban areas turn these same teeth (and antlers, in mating season, when bucks are rubbing the velvet off their antlers) onto landscape plants as well, and will raid gardens as well as field crops. There are also at least 1.5 million collisions between deer and vehicles every year, killing more than 100 people and injuring tens of thousands. And, deer are very hard to control in urban and suburban areas, where ordinary hunting is out of the question, and authorities may be reduced to trapping them and killing them with a bolt gun, slaughterhouse style, to reduce their numbers.

Now, as before Europeans arrived in this country, coyotes coexist successfully with many wild ungulates. Photo by D. J. Hannigan

A return to a more natural situation?

From the time wolves and other large predators were exterminated from the East, and for that matter largely from the West, the deer herds had no real predator, except man, to keep their numbers in check. What hunters today are really complaining about is a return to a more natural situation, where there is another predator—the coyote—that they must share the deer with. Still, as noted, except in a few limited situations, there is no real scarcity of deer, and in fact there are fewer American hunters today than ever before to compete with each other. Hunters in this country in 2010 number 11 million, as opposed to more than 19 million in 1975. But even when the deer in a particular area are actually overpopulated, prospective hunters are likely to feel there are not enough deer.

If you do need to reduce coyote predation on deer

Here are a few things that will help. First, if you are going to remove coyotes from an area for this reason, do so after they establish breeding pairs, and before the fawning season—this means in very late winter or early spring.

If you have a farm or other rural property, encourage the growth of thick vegetation near woodlots that will make good cover for fawns. Don't link deer food plots and other feeding sites with trails; coyotes love to follow trails and it will encourage their movement into key deer areas.

I'VE HEARD THAT COYOTES ARE REALLY HARD ON LIVESTOCK (KILL A LOT OF IT). IS THAT TRUE?

For more than 150 years, this has been the most hotly contested question in the whole universe of finger-pointing at the coyote. This is the belief that caused the U.S. government to embark upon, and continue even to this day, its campaign of coyote killing, and stock-raisers and environmentalists have battled over it endlessly. As Hope Ryden put it at the height of the controversy, "the environmentalists are in one camp and sheepmen and their 'hoofed locusts' on the other, and neither side trusts, or even really communicates with, the other."

As the dust has settled at least a little on this issue, the modern view is heading toward the less inflammatory idea that coyotes definitely do prey on livestock, sometimes significantly, but in general they are not the rapacious stock-killers they have often been made out to be. Domestic dogs, for instance, pets running loose, or feral packs, do almost as much damage to livestock, and sometimes more. Some, but by no means all, coyotes do become livestock killers. But the help that coyotes give ranchers and farmers by reducing many of the animals, such as rabbits and rodents, that feed on their crops and pastures almost offsets this.

We humans have also contributed to the problem by our endless tinkering with the genetics of our livestock, creating many breeds of animals that are very docile, gain weight fast, and have other habits and traits we want them to have. For the most part this also makes them extremely susceptible to predation because they don't run fast and are not super alert, nor do they exercise parental protection of their young to the degree many wild animals do.

Wily the sheep killer

Sheep are the livestock coyotes are most clearly implicated in preying on. In 2010, for instance, predators accounted for 34% of sheep and lamb deaths in Montana, and 78% of those animals died under the teeth of coyotes. For lambs, nationwide, coyotes were the number-one cause of death, more than even harsh weather or disease.

It's easy to see why sheep and their young ones are the most frequent target of coyote attacks. Sheep have been domesticated for so long that they are renowned for their helplessness, are a size that coyotes can handle, and are often out on the range or other pastures where no one is watching. They also taste good, as leg of lamb lovers would attest.

Which coyotes are most likely to be offenders?

The most likely offenders are usually the dominant animals in a territory, the alpha pair, usually a two-year-old (or older) male and his mate, especially when they are raising young.

When are sheep most vulnerable?

When they have small, largely defenseless young ones, of course, but also at key points in the coyote breeding cycle—when coyotes have pups to feed, and when they are teaching those pups to hunt for a living. In most parts of the country this means spring through late summer, plus perhaps late winter, when prey of any kind can be scarce for coyotes.

Interestingly, some scientists say that at any time of year, more sheep are killed by coyotes on foggy or rainy days.

How are they attacked?

Usually, coyotes approach a flock and circle it. When a sheep breaks away from the rest, coyotes run it down and attack it. Most of the time, they do so by biting the sheep in the throat behind the ear, then bracing their feet to stop it from running. Then the coyotes shift their jaws to get a firm grip in the area of the larynx, and hold on until the sheep suffocates, or falls to the ground. They start eating as soon as the sheep stops struggling. Lambs are usually bitten on the head or spine. Whether they have attacked an adult sheep or a lamb, coyotes may add here to their bad reputation for eating while the animal in question is still alive, and in this case bleating.

Coyotes, unlike dogs, don't kill for the fun of it. They usually kill to eat. But occasionally, when attacking sheep or other livestock, they may go on what seems to be a killing spree, killing up to a dozen animals in a single night. The experts think this is because they are either teaching their young ones to hunt, or get caught up in the inbuilt predator-prey response (that is, when an animal runs, chase it and kill it).

Ways to protect sheep

Determined farmers, ranchers, and researchers have come up with a whole battery of possible ways to protect sheep from coyotes, and here is the "short form" of many of them.

CHOOSE BETTER BREEDS Focus on breeds such as those with stronger flocking instincts.

INSTALL EFFECTIVE FENCING For information on how to accomplish this, see pages 121–127.

KEEP CLOSER WATCH ON THE FLOCK In the old days, this would have been done by full-time shepherds. Since for the most part this sound practice has been discontinued for cost, or lack of employees willing to do it, this means checking your flocks yourself on an irregular schedule, day and night, and moving them closer to human contact. Especially avoid remote pastures, or those near rugged cover, ravines or gullies, or watercourses.

REMOVE COYOTE COVER Remove heavy brush, stump piles, and junk piles that harbor rabbits and rodents that attract coyotes—from pasture areas.

PUT SHEEP IN BARNS, SHEDS, OR NEARBY CORRALS AT NIGHT They quickly learn to appear at the right time for you to do this, as if they appreciate the greater protection from both predators and weather this gives them.

Have ewes do their lambing in such places. Labor and feed costs will be higher, but more lambs will survive to go to market. You can also concentrate the lambing period, so that highly vulnerable youngsters are part of the flock for a shorter span.

GET A GUARD ANIMAL For information on choosing a guard animal, see pages 82–83.

USE COYOTE FRIGHTENING OR DISCOURAGING DEVICES These are described on pages 129–130. Lighting corrals at night is particularly helpful. Sometimes parking a truck or car in a pasture can aid the cause (and if it doesn't scare the coyotes off, it makes a good blind to shoot them from).

LIVESTOCK PROTECTION COLLARS Since sheep are so often attacked at the neck, a variety of livestock protection collars have been developed. The basic idea here is that if a coyote grabs a sheep by the throat, he will encounter a nasty surprise—anything from hot pepper juice to electric shocks to the deadly poison compound 1080 or cyanide. Collars like this are expensive, call for special training before you can use them, and the sheep wearing them is usually killed anyway. Most important of all, the bugs on all of them have not fully been worked out, as of this writing.

CLEAN UP Do not leave the bodies of sheep or lambs that have died of disease or whatever lying about, or toss them in an open pit. Bury them at least two feet down, burn them, or have them hauled away. Studies have shown that feeding on dead animals can lead to feeding on live, and carrion like this attracts coyotes from quite a distance to your property.

KEEP GOOD RECORDS OF YOUR STOCK The coyote sheep predation problem is worsened by the fact that many ranchers and farmers don't really know, until a situation is well advanced, that they are even losing sheep to predators, and when. This also makes it difficult to pinpoint the predator responsible.

KEEP CATTLE IN THE PASTURE WITH SHEEP (OR GOATS) This does seem to repel coyotes to some extent.

OTHER EXPERIMENTAL IDEAS These are ideas that have not fully worked yet, but have been tried, and may be better implemented in the future. One of these is aversion conditioning: leaving out baits with nonlethal chemicals in them that will cause coyotes to get sick after eating them, and hopefully avoid that food source in the future. (Coyotes do seem to have an effective but mysterious way of warning one another off of things that have proven to be unsafe to snack on.) Another idea is to spray livestock with something whose odor repels coyotes; this is still in the study mode because coyotes primarily hunt by sight, at least at first. And last but not least, birth control for wild coyotes is still in the unsatisfactory experimental stages.

Ways of totally excluding predators from sheep are also under study or in early implementation. This means raising sheep and lambs in total confinement, as much poultry and some pigs are now raised in this country. Although this is not an attractive alternative from the point of view of the quality of life of the livestock in question, it certainly is a way, for the most part, to exclude predators.

CHANGE YOUR ATTITUDE As Shreve Stockton points out in *The Daily Coyote,* maybe those who raise livestock, especially on government-owned ranges, should accept a certain amount of losses to predators of all kinds as "the nature tax."

What about goats? Are coyotes a threat to them?

Coyotes like goats as well as, or maybe even more than, sheep, and the statistics of yearly loss to coyotes, now that goats are such a popular livestock animal, support this. Angora goats are particularly susceptible. The same protection strategies outlined above apply to goats, which are in the size range of sheep.

Are pigs at risk from coyotes?

They are—piglets and young pigs, for the most part. Coyotes may sneak in and snatch piglets from their bed with the mother, or even right from the teat. Pigs up to about ten pounds may also be found partly eaten, often killed by a bite to the head. Failing to dispose quickly of the carcasses of dead pigs will add greatly to any pig-snatching problems. See pages 121–127 for coyote-resistant fencing measures for hog lots, or consider a donkey as a guard animal (discussed later in this chapter).

How about poultry?

Every predator and varmint around pursues poultry, and coyotes are no exception. In 1990, for instance, 17,000 domestic poultry were killed by coyotes, especially turkeys. For the home poultry raiser, the best defenses are 1) not letting chickens and the like run free; 2) having a tall, sturdy fence around the poultry yard; and 3) shutting the poultry house *without fail,* every day, before dark.

How much of a problem are coyotes for cattle?

Coyote predation is much less of an issue with cattle. Coyotes roam many cow pastures without doing any harm, simply feeding on the afterbirth that emerges when a calf is born. When coyotes attack cattle—some coyotes learn to do this and then teach their young—it is mainly young calves that are targeted, newly born or less than two months old. The National Agricultural Statistics Service reported 95,000 calves across the country were lost to coyotes in the year 2000.

A mother cow defending a calf can be a formidable opponent, so coyotes mostly take the offspring of inexperienced, first-time mothers, called heifers. This usually happens when the cow leaves the calf for a while to feed and water herself. Occasionally, coyotes may attack a cow that is having problems in the birth process. Approaching the exhausted and weakened animal, they may start feeding on the emerging calf—and perhaps the hindquarters of the mother as well.

Is it true that a tailless calf is often the work of coyotes?

Some coyotes (and for that matter, some feral dogs) do get into the habit of snipping the tail, or about half the tail, off a young calf and eating it. It is thought that this comes about because coyotes like to lick off the manure that clings to the rears of very young calves, rich with the heavy milk a cow gives right after birth, called colostrum. It is not too much of

Photo by D. J. Hannigan, law enforcement, Park County, Colorado

a step from that, to taking a little bite of the calf itself, especially if the calf happens to incur a nip or scratch in the process. Bobtailed or tailless calves usually survive the operation fine, though some cattle raisers complain that they cannot be sold to feedlots in northern states, since this leaves their rear ends unprotected.

Cattle protection measures

- Keep a close eye on first-time mothers, and those who have proven themselves to be mediocre parents. Bring cows about to calve closer to the farmstead if you can, and try to prevent cows in pastures from gaining access to the kinds of secluded locations they will pick to have their calves, if given the chance.
- Install coyote-resistant fencing (see pages 121–127).

- Plan breeding to shorten the number of weeks during which the herd will contain small, helpless youngsters.
- For heifers, consider using bulls, called calving-ease bulls, that will lessen the chance of oversize offspring and hence birth difficulties.
- Avoid discarding carcasses that encourage predation, as noted earlier.
- Get a guard animal.

GUARD ANIMALS: THE WAVE OF THE FUTURE IN LIVESTOCK PROTECTION

They're not a cure for all of the problems of protecting stock in field or pasture, but especially when combined with good fencing, guard animals go a long way toward giving a livestock owner some peace of mind. As Cheri VanderSluis of the Maple Farm Sanctuary in Mendon, Massachusetts, explains, "Since we have had our Maremma livestock guard dogs to protect our smaller farm animals, we haven't had one loss to coyote predation in 15 years. We can now appreciate the coyote's beauty and grace."

Guard Dogs

Dogs are perhaps the most common guard animals, and we are not talking here about Australian shepherds or border collies, the kinds of dogs used to move sheep and other animals about. Livestock guard dogs are for the most part very large, independent-minded animals whose breeds have a long history of defending the flocks they are given responsibility for. They are mostly white, like the sheep they often guard, and have exotic names like Maremma, Komondor, Akbash, Kangal, and Great Pyrenees because they for the most part originated in other countries.

Dogs meant for guard duty are put in with young livestock for a couple of months when they themselves are only seven or eight weeks old, to bond with their charges. They are given some basic obedience training, but not made into household or farmstead pets. Guardian dogs work best in small pastures with good visibility in all directions, and one mature and experienced dog can tend more than a hundred acres, containing hundreds of sheep. Guard dogs sometimes show no knack for guarding (some breeders have a money-back guarantee for this), or may chase wildlife or even people they don't know, and occasionally play too rough with their charges. But they are a poison-free, trap-free, firearms-free alternative. Sites such as **canismajor.com/dog/livestock** and **nal.usda.gov/awic/companimals/guarddogs/guarddogs.htm** give a more complete list of guard dog breeds and more details, and there are also entire books on this subject.

Is it true that donkeys can help scare off coyotes too?

They can. Donkeys have an instinctual dislike for doglike animals, including coyotes, and will attack them, or drive them from an area, given the chance. They will run at any intruding canid, braying and baring their teeth, biting and kicking. Donkeys live longer than guard dogs, and often cost less—many people buy the feral burros sold by the government for this purpose. They can also graze for most of their food right along with their charges, and need only an occasional worming, hoof trimming, and vaccination. One donkey can look after 150 to 200 sheep, but you have to make sure there is no other donkey or horse in with them, or near them. Otherwise they will ignore the sheep in favor of equine company. Only jennies (female donkeys) or gelded males are used for guarding.

How about llamas?

Those tall, commanding creatures, which can weigh up to 500 pounds, make very effective coyote-chasers, too. Like donkeys, they live longer than dogs, have an inbuilt dislike for canids, need no special food and little training, and overall are thought to perform even better than donkeys as guard animals. They not only try to drive off attacking canids, they make loud alarm cries and herd their charges to a safer area in the case of an attack. Even the llama's smaller relative, the alpaca, gets good marks for livestock protection. One gelded male llama can guard 250 to 300 sheep.

A COYOTE DID IT . . . OR DIDN'T . . .

When we find dead animals or birds, or what is left of them, is there any way of knowing whether or not they were the victims of a coyote attack? This is not a simple matter (there are many predators besides coyotes, and coyotes are scavengers as well as hunters)—it can befuddle even the experts at times. But there are definitely clues, and you may be able to find and decipher them.

First of all, though you may be angry or upset, it is important to take a good look at the "crime scene" as quickly as possible. Once time has passed, and other scavengers as well as weather changes have entered the picture, finding useful evidence becomes much harder if not impossible.

Examine the entire area carefully. Are there tracks or droppings anywhere? What kind? (See pages 36–38.) Are there signs of a struggle, such as blood on the ground or scattered around, mashed or broken vegetation, torn up earth or grass? Blood and disturbed terrain usually mean a predator, not death from other causes. Is there hair caught in nearby fences? Coyote hair is usually black mixed with tan, or banded, like hair from a striped cat.

Is one animal or bird killed, or more than one? Have they been eaten, or merely killed? Is there just one body, or traces of it, near the edge of a field or other cover, or are there victims scattered all over? What parts have been eaten, if the animal has been fed upon? Are the remaining animals upset and nervous? Check the following chart for the meaning of these things.

Aside from blood in the area, there are some other clear indications of death by predator attack, though you have to be willing to do a little dissecting to find them. If a predator killed a sheep or other animal, there will be tooth marks, and under the skin where those marks are, bruises and hemorrhages. This means the animal was bitten while it was still alive.

If the animal in question is a lamb, if it was born alive and then killed by a predator, its stomach will contain milk, and the stump of its umbilical cord will have a blood clot on the end. The hooves of lambs are covered with a thin tissue when they are born, and if they were born live and stood on the ground, that tissue will be worn off, and there may be dirt on the hooves. The lungs of lambs born live will be pink and spongy; if the lamb was stillborn, the lungs will be dark red.

WHODUNIT CLUES: Dogs versus coyotes

Since dogs, pet or feral, have been known to kill as much or more livestock as coyotes, let's do the dog–coyote shakedown first.

Number of animals or birds killed

Though coyotes do occasionally make multiple kills, dogs are far more likely to kill everything they can catch, or everything in sight.

Where the body is located

Coyotes usually make their kills from or near cover, such as a fencerow. The victims of a dog attack will be scattered all over, in plain view. Dogs leave their victims where they fell, whereas coyotes will often drag a carcass away as they feed on it.

Is the animal or bird eaten?

Coyotes do not for the most part waste their prey—they eat it. Dogs, with the possible exception of feral dogs, rarely do.

Where are the wounds on the animal?

Although inexperienced coyotes and pups will sometimes attack an animal wherever they can, coyotes normally kill sheep by grabbing them by the neck, just behind the jaw and below the ear. A sheep attacked by coyotes will have puncture wounds on the neck, bruises beneath the neck skin, and perhaps some foamy blood by the windpipe. Coyotes may attack deer too at the throat, or from behind, at the rear and flanks.

Animals attacked by dogs will usually have wounds all over, especially on the rear part of the body, and on the head and face, and may be merely mutilated rather than killed.

How are the other animals?

Dog attacks, which are usually extended and noisy, stress the other animals in a herd or flock. Coyotes just kill one animal or bird quietly, so the rest of the herd may scarcely notice it. Dogs often chase cattle, sheep, and deer until they are exhausted and overheated; coyotes rarely do.

PREDATOR PROFILES: Coyotes versus other wild predators

COYOTES Usually kill smaller animals by a bite to the back of the head, or neck. They normally kill sheep, as noted above, by biting the neck or throat. In winter, when sheep have heavy wool in this area, they may attack the hindquarters. Puncture marks from a coyote's teeth will be $1\frac{1}{8}$ to $1\frac{3}{8}$ inches apart in the case of the upper canines, and 1 to $1\frac{1}{4}$ inches for the lower.

Coyotes are usually most interested in the afterbirth, but they may attack cows, especially first-time mothers, while they are giving birth, which results in serious injury to the hindquarters of the cow as well as the death of the calf. They may attack larger calves at the hindquarters, or hamstring them. If the animal was pulled down by the head, as sometimes happens, it may have a broken jaw or wounds on the face. Coyotes will sometimes eat the tails off of live calves.

Deer may be attacked at the hindquarters, or the neck.

When coyotes feed on a large animal, they usually begin behind the ribs, and eat the internal organs such as the liver, heart, lungs, and kidneys first. They will eat muscle meat next and finally may even drag off and eat many of the bones. In the end only part of the hide and part of the skeleton are left. They will occasionally cache a kill to continue feeding on it later.

Small animals, such as pets or small lambs, may be carried off out of sight, or dismembered with a coyote's sharp teeth to carry to the den to feed pups. Of animals like rabbits or fawns there may only be scattered fur and a few bits of bone left.

BEARS May make multiple kills and usually kill larger prey by crushing the skull, neck bones, or spine, or with a powerful blow to the head or shoulders. In a bear kill there will often be large claw marks on the neck and back or elsewhere on the victim. Bears will eat the udder of a nursing ewe first, prefer muscle meat to entrails, and often partially skin their prey. Bears often cover a partially consumed carcass to return to it later.

BOBCATS Jump on the back of larger animals, and seize the throat, neck, or head of smaller ones. There will be claw marks on the neck and back of the prey animal, and tooth marks that are ¾ of an inch to 1 inch apart. Bobcats are more likely to eat the stomach and intestines than coyotes are. Bobcats may cover a carcass with brush or dirt.

COUGARS Usually attack their prey's head, neck, or back, leaving large claw marks, and often a broken neck on the animal. Cougars have big teeth, too: their upper canines are 1¾ to 2½ inches apart, lower canines 1¼to 1¾ inches apart. There may be drag marks, and the carcass is often covered with soil, grass, or snow. Cougars may make multiple kills.

WOLVES The punctures from a wolf's tooth are ¼ inch in diameter; a coyote's ⅛ inch or less. Wolf bites also create more severe bruising and tissue damage under the skin than coyote bites do. The powerful jaws of a wolf grab an animal just once at the throat; a coyote often bites the throat several times as it switches positions to maintain its grip. Wolves always eat what they kill, and often attack a large animal from the rear.

FOXES Their canine teeth are smaller and closer together than a coyote's: just ½ to ¾ of an inch apart for gray foxes and 1 inch to $1\frac{1}{16}$ inches apart for red foxes. They often attack at the neck as a coyote does, but the prey animal often ends up full of bite and scratch marks, as the smaller fox has more trouble subduing it. Foxes will eat the internal organs first, and may eat the nose or tongue. They also may partially bury an animal near where it was killed.

EAGLES Deep punctures from an eagle's talons at the top of the head, neck, or back may be 1 to 3 inches apart. Animals may be skinned, and there are often white bird droppings at the scene. Eagles sometimes eat the brains of animals they kill.

Poultry predators

Just about everything loves chicken (and ducks, geese, turkeys, and such), so you have to pay close attention to the clues here.

COYOTES Carry off one bird at a time, usually at night, often by digging under the fence. It has often been noted that coyotes have a very piercing look in their eyes, and an old Mexican legend insists that the "electric" stare of a coyote can even charm chickens and other poultry out of their roosts in trees. Another long-established belief here is that coyotes always grab a chicken by the neck so that it can't give a distress call.

FOXES Kill one bird at a time at night, often carrying it off. There will often be more feathers scattered around when a fox abducts poultry than when a coyote does.

HAWKS One bird disappears during the day, or is killed in the yard and plucked before it is fed upon. Hawks are likely to be spotted returning another day for seconds. Chickens and ducks will make distress calls when a hawk is in or near their pen.

OWLS Such as the Great Horned Owl, a huge, common, and formidable night hunter, often kill one bird nightly. They often eat the head and/or neck, which they pick clean neatly.

MINKS AND WEASELS May kill multiple birds by bites to the head or neck; none may be eaten.

RACCOONS Often remove the head of a bird and eat the crop and breast. Large coons can kill even birds as large as adult geese, and full-grown cats. Raccoons can reach their hands into a cage or fence made of wire mesh, and kill birds by tearing off their heads or legs.

OPOSSUMS AND SKUNKS Are more likely to maim adult poultry than kill it. They are fond of eggs.

RATS Will grab baby poultry or eggs and carry them to their burrow.

DOGS Will kill a number of birds and leave them scattered around but not eaten.

WHAT IF I HAVE NO CHOICE BUT TO ATTEMPT TO REMOVE STOCK-KILLING COYOTES?

When livestock predation is a persistent problem in a particular area, the best solution may finally be to remove the offending coyotes. This means not just reducing the number of coyotes in the area, but trying to identify and remove the specific ones that are doing the damage. See Chapter 7 for a concise guide to coyote hunting and trapping, two of the main ways of eliminating repeat stock killers.

The arsenal of poisons once used indiscriminately on coyotes has now been narrowed down to only one that can be used, and that only in some areas, called the M-44. This is a device that gives a coyote a mouthful of sodium cyanide when he pulls the spring-loaded tab treated with scents meant to tempt him into doing just that. This is not the best way to target specific animals, even where legal.

Studies have shown that it is the alpha pair in an area, a mature, usually two-year-old male and his mate, that are the mostly likely to kill livestock. The best time to remove them is at the beginning of the breeding season, in late winter or very early spring. Removing coyotes when they are often hunted, in fall and winter, may actually increase the number of coyotes in an area by assuring the survivors better nutrition for the breeding season and those hungry pups to come.

Coyotes are a common sight in suburbia and cities today.
Photo by Jonathan Way, www.easterncoyoteresearch.com

CHAPTER 5
the city coyote

About 20 years ago, starting in the early 1990s, a new chapter began in the dramatic story of the coyote takeover of America. Coyotes stopped limiting themselves to farms and forests and began to find their homes and dens right in the midst of suburbia, and even our cities themselves. From Albany to Houston, Milwaukee to Washington, D.C., Cincinnati to Santa Barbara, there was a new urban resident on the block, delighting some people and making others uneasy. This chapter takes a good look at the urban coyote, and at the concerns many people have about their increasing aggressiveness to pets and people.

COYOTE TRACKS IN CENTRAL PARK: THE RISE OF THE URBAN COYOTE

> *"The ghost of the plains has become the ghost of the cities."*
> —Stanley D. Gehrt, The Ohio State University urban coyote researcher

In recent years coyotes have been sighted, to the delight of the media, in the heart of Manhattan—Central Park, and in Harlem, Queens, and

Brooklyn, at Columbia University, and right beside the Holland Tunnel. The gray and tan ghost also has been making himself at home in the heart of downtown Detroit, Philadelphia, Atlanta, St. Louis, Nashville, Cleveland, and Phoenix. Coyotes have blundered into a sandwich shop in Chicago, a furniture store in Kansas City, the elevator of an office building in Seattle, a house in Hatfield, Massachusetts, and a toll plaza near Boston, among many other places. By 2010, it was estimated that there were 5,000 coyotes in Los Angeles, and at least 2,000 in greater Chicago, one of the most urbanized areas in the country. Once they really started looking, said coyote researcher Gehrt, he and his fellow researchers couldn't find anywhere in Chicago where coyotes weren't. Was the occupation of America's suburbs and cities just the obvious next step in the awesome range expansion of the coyote in the past 100 years, or, as some people believe, the result of urbanization that has pushed animals and birds of all kinds across the country into new circumstances—or out of existence? No one is quite sure, but like the rat, starling, raccoon, opossum, and skunk, the coyote has found life close to humans entirely possible—in fact, attractive.

Suburban and urban areas offer a rich array of foods that are not hard to catch or come by, and no scarcity of cover, either, admittedly of sometimes quite a different kind. The push to add attractive plantings to streets and highways has only added to the prey and cover possibilities. Most cities were built on or near major waterways, and have plenty of swimming pools and pet water dishes, so water is not a problem, either. The really surprising thing is that an animal as large as a coyote—and a predator at that—has been able to squeeze himself into this new landscape, and prosper. And prosper they have. City coyotes not only live longer than rural coyotes, but there are more of them, three to six times more, in every square mile of urban areas, than out where the buffalo used to roam, or other wilder parts of the country. When the authorities in Glendale, California, decided to remove all of the coyotes near where a child was attacked, they found themselves removing more than 80 animals.

WHERE DO URBAN COYOTES ACTUALLY LIVE?

Given the bountiful supply of edibles in suburbia, urban coyotes for the most part have smaller territories than their rural relatives. Solitary coyotes in metropolitan areas, as in wilder regions, have a larger range, up to 25 square miles or even more, but pack members can get by nicely in the burbs with a home range of two to ten square miles, usually at the lower end of that figure. Coyotes in some parts of Los Angeles have territories of less than half a mile. Foxes and coons who take up city life also have smaller territories than elsewhere, we note.

To make up their home range, urban coyotes piece together a patchwork of which the most important parts, usually, are areas that on a small scale resemble the prairies, fields, and woods they left behind—parks of all sizes, golf courses, cemeteries, vacant lots, woodlots, baseball fields, airports, military bases, reservoir watersheds, and forest preserves. Not even the smallest strip of vegetation or open land is neglected, so coyotes also make use of roadsides and highway rights of way, power-line corridors, land around train tracks, dumps and landfills, the green space and small ponds (with the usual ring of reeds and cattails) of industrial parks, plus of course the lawns, yards, and gardens of private homes. Not neglected, either, are the "leftovers" that many people pay little attention to, like swamps and gullies and washes, the latter of which make excellent routes for invisible travel. Many urban coyotes also include bits of the most citified surroundings imaginable in their territory, along with these mini-greenbelts. One urban pack in Chicago, for instance, lives at the intersection of two interstate highways, near the second largest indoor mall in the United States.

For shelter, when urban coyotes can't find the bushes or tall grass they might find elsewhere to bed down in for the day, they may curl up underneath porches or in abandoned buildings, or in a patch of shrubs not far from a house. And urban coyote dens can be in anything from large dry drainpipes and storm drains, to junked cars to within the mausoleums of cemeteries. They also den beneath garden sheds or tool sheds (although their favorite urban denning place is city parks).

To use most of these places unobtrusively, however, does require that urban coyotes be even more nocturnal than their country cousins, to avoid people as much as possible, and preferably avoid even being seen by them. There is also less traffic at night, which is good, because being run over by cars and trucks, as you might expect, is the number-one killer of urban coyotes. Most urban coyotes have to cross many roads in their night's hunting, and some of those roads are traveled by more than 100,000 vehicles a day. One UPS driver reported seeing a coyote cross eight lanes of heavy traffic of an interstate highway in northern Kentucky, right in the middle of the day, and reach the other side safely.

Urban and suburban coyotes cross many roads every day and night, and they do not always survive the experience. Courtesy of Colorado Division of Wildlife (Michael Seraphin)

Is it true that coyotes like airports?

Airports, with their nice, big, open spaces and expanses of grass, are one more urban and suburban landscape that attracts coyotes. There are not many competing predators out there, and plenty of rodents and rabbits. A pilot from Dallas in the 1980s reported that there were coyotes on the runway when he left Texas, and a coyote on the runway to greet him when he arrived in Dayton, Ohio! In 2010, when a coyote at Portland International Airport was chased off the runway, he darted under several trains and then boarded the light-rail train headed for downtown, taking a window seat. (Airport officials did manage to remove him peacefully to a more appropriate part of their property.)

Some smaller airports, such as the one in Middletown, Ohio, have actually welcomed coyotes on their premises, because they reduce or eliminate Canada geese, which endanger planes and make a big mess with their wet and slippery green droppings. But most airports have installed tall fences, when necessary, to keep coyotes out, because coyotes on the ground can be hit by the landing gear of small or large planes, and even occasionally be thrown into the air and sucked into a jet engine. They sometimes make other mischief, too, such as chewing through those ultra-expensive fiber-optic cables.

WHAT DO URBAN COYOTES EAT?

What are urban coyotes looking for in their nightly forays, which can take them through a number of adjoining towns? Rodents is one big answer—no one is unhappy to know that they spend much of their time catching mice of all kinds, and rats. It's often been estimated that there are as many rats in the average city as there are people, and full-grown rats are too large and aggressive for many cats to take on. There are more squirrels, chipmunks, gophers, raccoons, and opossums per acre in the average suburb than in the same amount of space in the wilds, and groundhogs, muskrats, and deer can be found in suburban landscapes, too. Deer are often all too plentiful in urban areas, with no hunting pressure on them.

In even the most developed areas, coyotes can usually find patches and preserves of open land for good hunting. A coyote hunting on the grounds of Fermi National Accelerator Laboratory in Batavia, Illinois. Photo by Brian E. Tang/Tang's Photo Memories

In recent years Canada geese have become not fully welcome year-round residents in many urban and suburban areas, fouling waterways and walkways with their droppings. Urban coyotes feasting on goose eggs have cut the yearly increase in city honkers back to more manageable proportions. Urban areas are full of bird feeders, too, with all of the small animals as well as birds they attract. Suburbs have gardens often left unharvested, compost piles full of ripe goodies, and landscape plants with edible fruits and seeds. Trash cans are yet another source of edibles, and by feeding on the all-too-plentiful roadkill, coyotes help keep roadsides more pleasant and attractive. Last but not least—an incendiary issue in coyote–human interactions, see pages 100 and 108—there are pets, cats and small dogs, plus the food often set outside for them.

A mature alpha male coyote eating a vole. Coyotes do a big service in urban areas, and elsewhere, by reducing the rodent population. Photo by Brian E. Tang/Tang's Photo Memories

THE HUMAN REACTION

The coyote occupation of the suburbs and cities was so quiet and low-key that at first few people were aware of it. Some people didn't even realize that what they were seeing wasn't dogs or wolves, and coyotes lived out their lives in many neighborhoods with residents oblivious to them. Then gradually the number of complaints about a strange animal in the neighborhood—coyotes—began to rise, from 20 a year in the early 1990s in the Windy City, for example, to more than 500 a year by the end of the decade. Researchers scrambled to study this new wild urban canid, which was not an easy job, since coyotes are notoriously hard to trap or, for that matter, observe.

One very positive result of it all was that many people, especially young people, are thrilled to have a touch of the wild on their doorstep. Something to make them feel, even in the midst of apartment complexes and high-rises, that they are still part of the natural world. As Catherine Reid noted, "I thrilled at the idea of a creature rising out of the very landscape we had sought to control with bulldozers and fences and streetlights and dams, something that could startle us with its teeth and resilience and a howl that punctured sleep and sped up breathing." Jonathan Way, PhD, one of the great champions of the urban coyote, said it well too: "One reason I admire coyotes is because they give a landscape, even an urban one, a sense of mystique . . . I am always a little more alert knowing that predators are living in an area."

On the other hand, some urban and suburban dwellers are made anxious by the very sight of a coyote. A lot of people in such areas call the authorities just because they have caught a glimpse of one. A number of states report "fear" as the primary complaint about coyotes, which is understandable, where people may not be accustomed to seeing or hearing them. As one blog posting noted, "We want to live in nicely landscaped and countrylike surroundings, but we don't want them to be the least bit dangerous or wild."

All of this makes for some very heated discussions. As Jim deVos, former research chief of the Arizona Game and Fish Department in Phoenix, noted: "We've had [civic] meetings where half the group sits on one side of the room and half on the other. One side wants them gone. They say, 'I'm scared to death of them. They're going to eat my children.' The other side says, 'Boy, that's cool! I got to see one!'"

A BIG BAD STEP IN COYOTE APPRECIATION: FEEDING THEM

Unfortunately, for many people appreciating wildlife leads to wanting to help it, and the first thing that usually comes to mind is to provide it with food. Feeding wildlife not only makes you feel like you've done something good, but then you get to watch the animals or birds feast on the bounty you've provided, usually right in sight. Thus, many urban people feed not just birds but squirrels, raccoons, deer, feral cats and dogs . . . and coyotes.

Feeding coyotes on purpose is the single surest way to convert them from interesting additions to the urban landscape, to problems and threats. Here is how it happens.

First of all, in some urban and suburban areas coyotes may be a little less afraid of humans to begin with, after generations of close exposure to this odd two-legged animal, and generations without being hunted, trapped, or otherwise pursued. We feed coyotes in many indirect ways as it is—see pages 130–132. Going one step further and intentionally feeding them makes coyotes—in case they missed the point earlier—directly associate humans with food. Feeding them just once will turn the tide. They will actively seek out food from human sources after that, become bolder and more aggressive, and perhaps tamer as well. All of these are bad things, and can eventually lead to attacks on pets, children, and maybe even adults, as discussed later in this chapter. Feeding coyotes is also a disaster for them in the end, as fed coyotes, such as the ones found begging along the road in national parks, are usually caught and put to death sooner or later. "A fed coyote is a dead coyote," as the experts say.

But people do not give up the wildlife-feeding impulse easily. As coyote researcher Stan Gehrt has noted, "When I tell people they need to stop [feeding coyotes], they get upset. Not just a little . . . we're talking irate. It will be interesting to see if we can change people's behavior before they change the coyotes' behavior."

Many places have ordinances against feeding wildlife, and if the coyote experts had their way, there would be serious penalties for feeding coyotes. As the coyote management guidelines developed by the University of California state, "Anyone who intentionally feeds coyotes is putting the entire neighborhood's pets and children at risk of coyote attack and serious injury." Or as another source puts it, is exposing their neighbors to physical and emotional trauma.

I'VE HEARD THAT COYOTES KILL CATS. IS THAT REALLY TRUE?

Not a single expert or information source about coyotes would dispute this—the only question is, exactly how often does this happen across the country?

Urban and suburban coyotes are the biggest offenders here, and until fairly recently, the highest percentage of house cat remains scientists found in coyote scat or stomachs and put in the published record was about 13%, as part of a study done in Washington State. Coyote researcher Stan Gehrt, who has headed a longstanding study of urban coyotes in Chicago, says that even in urban and suburban settings, coyotes predominantly eat traditional coyote foods such as small rodents, rabbits, fruit, and white-tail deer. But reports of attacks on cats in almost every state have been increasing, and studies during this decade done in Claremont, California, Seattle, and Vancouver indicate that urban coyotes rely on pets as a major food source, especially in winter and spring. From 1985 to 1995 alone, the number of attack on pets in Texas rose fourfold.

Shannon Grubbs of the University of Arizona and Paul Krausman of the University of Montana took a closer look at the cat–coyote question in late 2005 and early 2006. They trapped just eight of the many coyotes in Tucson, radio-collared them, and observed their interactions with cats closely. The bottom line was that of the 36 coyote–cat encounters they witnessed, 19 ended with a dead cat. (The majority of the other 17 meetings involved what amounted to unsuccessful attempts by the coyote to kill the cat, such as by chasing it or lunging at it.) Furthermore, during the 45 times they happened to actually observe an urban coyote eating something in the course of their research, more than 40% of the time it was a cat. To quote one of the authors of this study, Paul Krausman, professor of Wildlife Conservation at the University of Montana, "The number of cats killed by coyotes in the West and nationwide is a lot higher than many people think."

No wonder we see so many "lost cat" posters and signs around.

Photo by Amber Pulse

The statistics brought to life

The cut-and-dried statistics above do not convey the pain, grief, and trauma people feel when a beloved pet—which is often considered a member of the family—disappears or is the victim of attack by coyotes. A woman in Atlanta reported on her experience in February 2010: "I heard a cat cry from the backyard, so I jumped out of bed and looked out the window, expecting to see another cat in our fenced yard trying to pick a fight with my nonagressive kitty. Two large coyotes were hovering over my precious little girl. I screamed and banged on the window and ran down to rescue her. When I got downstairs, the coyotes were still in the yard, so I banged on the window of the door and flung it open quickly. They jumped over the five-foot fence around our backyard easily and ran away. My kitty was still breathing, but had sustained severe damage to her head and passed away on the way to the ER vet clinic. This will haunt me for a long time."

In Chatsworth, California, coyotes came into a yard in 2001 and took a pet cat out of the hands of a 19-month-old toddler. In Oildale, California, a year earlier, a pair of coyotes treed a woman's cat, and then turned on her when she tried to protect it. In the summer of 1999, reports nuisance animal expert John Trout Jr., 55 cat collars were found in a coyote den on the outskirts of Surrey, British Columbia. In June 2010, a woman and her husband in Gulf Breeze, Florida, awoke early one morning to see a pair of growling coyotes playing tug-of-war with a neighbor's cat.

Why would coyotes prey on cats?

The most common answer to this question from scientists is that "coyotes sometimes kill pet and feral cats for food, or to remove competition [the competition of a smaller predator in the same area]." There are other answers, too, such as the fact that cats are not much bigger than rabbits, one of the great staples of the coyote diet. Then too, coyotes have no way of knowing that these particular medium-sized furry creatures are special to us—to them cats seem like just another prey species on the landscape. Plus in urban and suburban areas, there are *lots* of cats, pet and stray.

Another reason, which has been mentioned by a number of authorities, but never really explained or substantiated in the popular press, is that coyotes are said to have a special attraction to cats. The naturalist Vernon Bailey was quoted by Frank Dobie in *The Voice of the Coyote* way back in 1947 as saying that the coyote "takes a special delight in killing cats." Other sources say, "coyotes relish cats as food items," or "coyotes know that one of their preferred foods, domestic cats, can be found where people live."

How do coyotes hunt cats?

Little has been observed or reported on this, but urban coyote expert Jonathan Way gives us one clue. "My coyote tracking has enabled me to observe, in a number of instances, coyotes trotting in a zigzag fashion through neighborhoods. I can only guess that they were catting," he says. (Other coyote experts note that coyotes hunt other prey in this manner as well.) Way also witnessed an incident in which one of the coyotes he tracked approached a cat in the classical stalking manner.

Coyotes usually kill animals the size of a cat by biting them in the neck or on the head. In the case of cats, a hold on the neck keeps the cat from being able to turn around and bite its attacker. One experienced coyote hunter noted another possible scenario, "Coyotes attack their prey quickly and decisively by grabbing its throat. This produces a quick kill and keeps the animal from making a sound. Then they run from the scene with the prey in their jaws." There are many reports of coyotes jumping a backyard fence, seizing a small pet, and then disappearing with it. Coyotes sometimes dismember their prey to take it back to the den to feed pups. A California report noted, "Municipal authorities and homeowners have sometimes found remains of dead house cats and mistakenly assumed they were mutilated by people practicing animal sacrifice, when in fact they were killed by coyotes."

Can cats defend themselves against coyotes?

Cats do have sharp claws, teeth, and the ability to climb trees, where coyotes for the most part can't follow. And cats have good night vision, and other sharp senses. But living the good life with humans has made many pet cats overweight, out of shape, and under-vigilant in the outside world of staying alive by fang and claw. Some cats that are accustomed to being outside, and many feral cats, have not lost the keen alertness and agility that would help them out here, but they are not necessarily in the majority. And even a strong, fit, and highly vigilant cat is up against a very fast and agile animal, in the case of a coyote, which is on average three or four times its size, and has a very impressive set of teeth. Cats have a little more of a chance against a lone coyote than a group of them.

The usual outcome of an attack

Few cats attacked by coyotes live to make it to veterinary emergency care. Dogs have a better chance of surviving to be patched up. There are a few exceptions, such as the cat reported in one online chat room to have survived a "kill-bite" wound to the throat. Or Buddy, a 17-pound bobtail cat, who battled a couple of coyotes that crept into his home in Kennewick, Washington, through a garage door accidentally left open. After his owner was awakened by "blood-curdling screams," Buddy was found undamaged, with coyote fur in his paws, and coyote fur and dung decorated the house.

Often when a cat is attacked or goes missing, there is little left to tell the tale except possibly a few tufts of fur. Other times, cats may be found "missing their insides," or with nothing but their head and legs left behind. "It isn't much fun to come out in the morning and find half of your cat on the front lawn," one pet owner noted. If there are remains of an attack, they may be some distance from where the attack originally occurred.

The making of a serial cat killer

Once coyotes get started killing cats and other small pets, the trouble is just beginning. As soon they discover this new food source, how abundant it is and how easily they can take advantage of it, they may well become specialists in exploiting it. Before long there may not be an outdoor cat left in the neighborhood. Worse yet, experienced pet killers may teach their pups the fine art of stalking pets, right along with, or even instead of, how to catch mice and rabbits.

Smart as they are, if coyotes realized what a hornet's nest of anger and desire for vengeance pet-killing stirred up, they would probably avoid it. But cats are usually killed when people aren't around to witness it and make their displeasure or fury known.

Other dangers to cats, and things they themselves endanger

All of this being said, coyotes are by no means the only, or worst, danger outdoor cats face. Cats are also at risk (great risk!) from the huge speeding hunks of metal known as vehicles, unfriendly dogs, foxes, raccoons, bobcats, fishers (large weasel-like animals of the northern U.S. and Canada), great horned owls, nasty humans, and deadly diseases they can contract from other cats in their travels.

And as bird lovers would be quick to point out, pet and feral cats themselves kill billions of songbirds, young game birds, and small mammals every year, including species that can ill afford this. For this reason and others, many people feel that the coyote's role in the reduction of the number of feral cats is actually a great blessing.

How can you protect your pet?

The ultimate way, all the experts say, is to make your pet an indoor cat. The idea of totally indoor cats has been "in" for some time now in cat lover and humane society circles, and it is often pointed out that indoor cats usually live at least twice as long as outside cats. This does mean a lot more scooping of litter boxes, and perhaps coping with some spraying indoors, even from neutered males and females, but it is the wave of the future. It may also be necessary for other reasons, in places where free-roaming pets of any kind are not allowed. As one veterinarian in favor of this approach pointed out, "I'm always amazed at how people can't imagine changing their pets' lives because they say it might affect their happiness. But I think it's better for your best friend to stay indoors and live a long life than to be eaten alive." When you are trying to convert an outdoor cat to indoors, it has been suggested that you try starting in winter, when it will be less eager to go outside anyway. It may take a month or more to achieve the conversion.

A compromise: Protected outdoors

Are you one of those who, wise as it may be for our animals, regrets the fact that their pets must give up the entire experience of the outdoors, with all of its rich sounds, sights, scents, and experiences? Worse yet, that they must surrender it to what in some parts of the country might be called an invasive species that has appeared on the urban scene and is now endangering them? There is a middle ground. You can fence your yard in one of the coyote-proofing ways noted on pages 121–127 or, better yet, build for your cat(s) a safe outdoor enclosure.

This means an enclosure that is as big as you can afford to make it, with a sturdy roof as well as sides. To protect the inhabitants from dogs as well as coyotes and other perils, the structure needs to have strong supports made of treated lumber or rot-resistant small logs like locust or cedar, and then an understructure of strong wire such as two-by-four welded wire, covered with smaller wire such as vinyl-coated chicken wire. Given coyotes' penchant for burrowing under fences, make sure that the enclosure is well reinforced where it meets the ground. You also will need a good cat door for access from indoors, and some nice landscaping (pet-safe plants, please!) plus a neat set of playthings inside—branches and logs to climb, stones to sit on, and a kitty platform or house or two, elevated if possible—cats always like the high view of things.

Check your local zoning laws and homeowner association rules while your cat enclosure is still in the planning stage, of course!

Pet owners who feed coyotes, intentionally or unintentionally, are asking for trouble. Photo by Steve and Dave Maslowski

Other ways to help protect your pets

If you can't afford to build a seriously coyote-resistant fence or enclosure, at least do the following:

- Follow the recommendations on pages 127–132 for making your property less attractive to coyotes in general, especially the tips involving pet food control; NEVER feed coyotes yourself or let your neighbors think it is a good idea to do so.
- Tempting as it is, do not feed feral cats. Both the food you put out and the cats themselves are at risk from Wily.
- Keep your pets in at night, at least (they could also be at risk in many areas in the daytime). This means shutting pet doors at night, too. Don't leave pets outside at night, even in a yard with one of the ordinary suburban kinds of fences.
- Should coyotes appear in your yard when you and your pet are there, pick your cat up immediately, if possible.
- If you lose a pet to coyotes, tell all the neighbors. Their cats are now at great risk, too.

Last thoughts on the subject

There are two sides to this story, one of which is eloquently summarized by the following two quotes.

In the foreword of *Suburban Howls* (Jonathan Way's book on the urban coyote), coyote expert Marc Bekoff, PhD, states: "Jon writes here that instead of worrying about losing our small pets to a coyote, we should be happy to realize that we live within a healthy ecosystem that is able to support a medium-sized predator like the coyote." And one of the many bulletins on coexisting with coyotes published by state governments across the country notes: "Coyotes get a bad rap. They're really fascinating animals. They do a lot of things that make people angry, like getting in the garbage or snatching up a cat, but they're just trying to survive. It's not a personal insult."

The other side is well expressed in Shreve Stockton's *The Daily Coyote,* a well-written account of Stockton's love affair with a pet coyote. She says, recording her feelings when "Charlie" was just a pup: "I was overcome with a visceral desire to destroy anything that might harm Charlie, to kill in order to assure Charlie's well-being . . . the desire to keep the animals that are in one's care safe from harm, from outside sources, is a powerful one."

ARE COYOTES DANGEROUS TO DOGS?

One might not think so. After all, dogs are members of the same biological family (Canidae) as coyotes, have many similar habits, and although dogs range in size from the two-pound Yorkie to the 200-pound Newfoundland, many of them are closer in size to the gray or tan ghost of the fields and suburbs. The bottom-line answer, however, is that at times coyotes will attack, and even kill, dogs of most kinds. Coyotes sometimes see dogs as prey, and sometimes as simple competition, but when either of these things happens the outcome is often not good.

To look at this by size category: The smallest of dogs—the toys, lap dogs, and miniature breeds (dogs up to say 15 pounds)—are most at risk, at all times of the year. In suburbs, cities, and, increasingly, rural areas, too, dogs in this size range are snatched from yards, lawns, parks, and play areas, and ever more often now, even from leashes while on a walk with their owners (more about this later). Consider just a few examples from recent headlines: A beloved Chihuahua is killed in a backyard in suburban Illinois; a small dog in Georgia left out by its owner to "do its business" is found later as a headless body; a Jack Russell terrier in Ohio needs three surgeries and 500 stitches after several coyotes go after it; a Maltese and Shih Tzu mix are killed in sight of their owner on Christmas Eve; and a miniature Schnauzer, and even a 24-pound Pekinese, are both killed by coyotes. Small dogs—and puppies of most breeds— are for the most part easy prey for coyotes, easy to overcome and easy to carry off, and especially attractive when coyotes have a family of pups to feed. A miniature dachshund is a lot easier to catch than a rabbit. Many of the small dogs attacked by coyotes, even if not spirited away, do not survive the encounter.

Medium-size dogs are somewhat less at risk, depending on the size of the nearby coyotes. But if a coyote wanders into the dog's yard—which the coyote will consider *his* yard—part of his coyote territory, a battle for ascendancy will follow. (Remember, dogs, wolves, and other canids always have a pecking order, as to who is on top of the doggie social ladder.) They may work the situation out nonviolently. But if the dog is smaller than the coyote, the coyote will expect it to give the dog-family signs of submission. If the dog fails to do so, or challenges it, the coyote (or group of coyotes) may well attack or even kill it.

Large dogs are the safest. Some large and swift breeds of dog are raised and trained specifically to kill coyotes, as explained in Chapter 7. Although coyotes can be fierce and effective fighters, they will usually avoid taking on the bigger breeds. In recent years, however, some dogs even as large and formidable as Rottweilers, Labradors, and pit bulls have been attacked by coyotes, even while accompanied by their masters.

Both large and medium-size dogs will run into trouble if they happen upon adult coyotes during mating season, especially near their den site, when pups are present. Coyotes' territorial instincts are intensified at such times, and they will immediately attempt to run off, or attack, the dog they see as a threat or intruder.

Dog protection measures

Not letting dogs run loose, even in official "natural areas" like state parks, can help prevent coyote–dog conflicts. Another prevention measure is never letting your dog "play" with coyotes. It may look like they are having fun at first, but if the dog makes a wrong move the coyotes could turn on it.

Should a den site be identified, be sure to keep your dogs (and other pets!) away. And since coyote parents are very sensitive to human intrusion around a den, you can usually do enough mild harassment, and making of your presence known, to cause the coyotes to move their family elsewhere.

On the home front, bear in mind that a fenced yard, alone, will not be much of an impediment to coyotes. Coyotes can easily leap over the average yard fence, and the fence may just make it easier for them to corner your dog. See pages 121–127 for what it takes to create a more reliably coyote-proof fence. Even if your dog is in a well-fenced yard, don't just stick it out there and forget it. Stay aware of what is happening out there.

Likewise, don't tie dogs outside and forget them. Keep dogs in a well-enclosed kennel (this means the top too!) if they must be outside at night.

The dog-walking dilemma

Taking your dog for a walk used to be just a chance for your pet to relieve itself, and for you to get a little exercise and fresh air; poop-scooping was about the only issue. In many parts of the country, suburbs and cities especially, walking the dog now requires more vigilance and fitness. Coyotes have figured out that a dog on a leash is extra-easy prey. Consider these dog-walkers' nightmares in the news across the nation: A 69-year-old man fights a coyote for seven minutes to save his 30-pound dog; a 51-year-old woman is bitten by a coyote while trying to protect her leashed Labrador; a woman falls and breaks her leg trying to keep a coyote from attacking her leashed dog; a man is bitten eight times trying to defend his dog from attack; a poodle is ripped from the arms of its owner; a coyote snatches a nine-pound mixed breed and breaks the leash to carry it off; a woman on a dog walk in town stops to check her cell phone messages and sees, out of the corner of her eye, a coyote headed straight for her dog.

Dogs have been injured and killed while being walked on leashes, and owners badly injured trying to prevent this, by coyotes that "came out of nowhere" in a flash.

Coyote-aware dog-walking

If coyotes are a problem for dog walkers in your area, consider the following protection measures.

- Carry a **coyote discourager** and be prepared to use it. This can be anything from a stout walking stick or cane, to a golf club, long pipe, or even a baseball bat. More militant types have even considered cattle prods. There are also spray repellents, such as Mace

Muzzle, specifically designed for doglike animals. (Mail carriers often carry these on their rounds for protection from menacing dogs, although experts are in disagreement as to exactly how effective they are.)

- Do not establish a regular routine for your walks: coyotes can learn it and lie in wait.
- Avoid walking small dogs, especially after dark, and any dog at dawn or dusk.
- Walk with another dog walker if possible.
- Stay in more populated areas, and away from good coyote hiding places such as thick shrubbery and woods.

(Yes, this could take some of the fun out of dog-walking!)

If a coyote approaches you and your dog:

- Scare it away.
- Pick up your dog immediately, if it is a small one. Don't wait until your pet is attacked; pick it up *before* that happens and head toward a safer place.
- Throw rocks, sticks, or whatever is at hand at the coyote.
- Stand tall and yell at it in a loud, authoritative voice.
- Don't drop to the ground to defend your pet—the coyote will see this as an act of submission.
- Use your coyote discourager if you have to!

WHAT ABOUT ATTACKS ON PEOPLE? WHAT CHANCE IS THERE THAT COYOTES WOULD ACTUALLY DO THAT?

This is probably the biggest question in people's minds when they ponder the fact that a medium-size predator with sharp teeth and claws is now living in our midst, in just about every suburb and city in the country.

The possibility of meeting an unfriendly coyote is a common concern today as coyotes get more used to humans and grow bolder. Photo by Steve and Dave Maslowski

The answer to this question would have been quite different 20 or even 10 years ago. The old answer was that coyote attacks on humans were exceedingly rare; now there is no doubt that they have been happening with increasing frequency. Between 1960 and 2006, there were at least 142 reported attacks on humans in the United States and Canada, according to a careful study of the subject done by Lynsey White and Stanley D. Gehrt of The Ohio State University. Southern California, which seems to be the testing ground of the urban coyote, leads the country in such statistics. The University of California, among other sources, says that there have been more than a hundred attacks on humans in that area alone since the 1970s. True, these are not big numbers, when we consider the other statistics coyote champions would quickly come forth with. Such as the fact that domestic dogs bite close to 5 million people a year in the United States, and about one-fifth of those bitten require medical attention. Or the fact that brown bears (the biggies, such as grizzlies) killed more than 300 people in the world during the 20th century, leopards more than 800, and tigers an astounding 12,000 plus. Or that our chances of being attacked by a coyote are smaller than the chance that we will be struck by lightning or attacked by a human predator.

However, the dog problem we brought on ourselves—we and our children live closely every day with more than 77 million pet dogs in this country. And we somehow tend to give little thought to dangers we have chosen to live with, such as the very large chance that we will be injured or killed in an auto accident.

This coyote attack issue is an entirely new one, and one that plays into one of our deepest leftover instinctual fears—the fear of being pursued and eaten by something.

As of mid-2010, only two people have been killed by coyotes. In 1981 three-year-old Kelly Keen of Glendale, California, wandered out the door without her mother's knowledge and was carried off by a coyote from the driveway of her home. She was rescued by her father but died despite four hours of surgery. She had a broken neck and had lost a great deal of blood.

In October 2009, a 19-year-old folksinger on tour took a break to hike in Cape Breton Highlands National Park, Nova Scotia. Taylor Mitchell chose the popular Skyline Trail, famous for its ocean views and even a chance to see moose. Attacked by two of the estimated 100 coyotes in this 366-square-mile park, she ended up on a Life Flight helicopter to Halifax, bite wounds covering her entire body. She died despite all efforts to save her. This was actually the third coyote attack on a human in this area of the park in recent years.

Though only two people have been killed by coyotes to date, experts have pointed out that more of the children who have been attacked would have died had it not been for the intervention of their parents—their vigorous efforts to drive off the attackers.

Children have been attacked by coyotes while playing in sandboxes or on swing sets, roller skating, snowboarding, eating cookies on their front porch, playing in their backyards, waiting to get onto merry-go-rounds, walking to school with siblings, eating lunch at school, walking with their parents or grandparents, and watching their parents play golf.

Adults have been attacked while mowing their lawns (on a riding mower!), picking up newspapers in their front yards, jogging, barbecuing, entertaining guests, sleeping on their lawns, lying on chaise lounges on their decks, sleeping in sleeping bags, talking on cell phones in their backyards, hiking, skiing, bicycling, and picnicking, hunting, doing home repairs, and attempting to take pictures of coyotes, even while stepping out for a smoke (a whole new reason to stop smoking!)

The trend is upward

Although more coyote attacks have occurred in the West than anywhere else, all across the country, more human–coyote encounters are being reported every year. As expert coyote hunter Tom Bechdel of Pennsylvania says, "If no attacks have happened in your state so far, you are lucky. It's only a matter of time." Scholarly sources echo this sentiment, as do many outdoor writers across the country. "The fatal attack in Cape Breton ensures that no one in Eastern Canada will look at coyotes the same way again," says author Don McLean. "It's too bad that it takes an innocent woman's agonizing death for people to see what [the coyote] is: an effective and opportunistic killing machine that will attack, kill, and eat whatever it can, whenever it can, wherever it can," says V. Paul Reynolds, Maine guide, author, and radio program host.

The coyotes that have been involved in most attacks on humans in recent years have not been rabid or handicapped animals (along the old story line that the leopard or elephant or whatever hunts humans does so because it has a broken paw or a rotted tusk). Nor have all of the attacking coyotes been willing to leave their intended victim or the area during or after the attack, despite frantic efforts by the people nearby to chase them off. And eastern coyotes are only more dangerous, because of their greater size.

The boldness progression

How did we get to this point? Urban and suburban coyotes have simply gotten bolder, and scientists have studied this phenomenon to the extent that they have established a clear-cut list of the stages coyotes go through until they reach the stage where they are a genuine danger to the inhabitants of an area. Here is that list.

1. Coyotes are seen more often in streets and yards at night.
2. Coyotes take pets at night and may approach humans.
3. Coyotes are seen in residential areas—streets, parks, and yards—in daylight, especially in the early morning and late afternoon. (In Southern California it was not unusual even ten years ago for newspaper deliverers, joggers, and other early risers to see one to six coyotes in the suburbs in a day.)
4. Coyotes chase and kill pets in the daytime.
5. Coyotes attack pets on leashes or near owners, and chase joggers, bicyclists, and other adults.
6. Coyotes are seen in and around schoolyards and other children's play areas, at midday.
7. Coyotes act aggressively toward adults in the daytime.

People in both urban and rural areas have noted the increase in coyote boldness in recent years. As one urban dweller noted, "They're so brazen now—they just stand there and look at you!"

The reasons behind it all

There are several good reasons for the increasing number of human–coyote clashes, and the new posture that many urban coyotes, especially, are taking toward people.

We are on the edge

As noted earlier, coyotes love "edge" habitat, and many suburbs and housing developments are right up against wild areas, or greenbelts. Many of the places in California that have had the worst problems with coyotes are housing areas that abut steep, brushy, mountainous habitat. A part of New Jersey that has had coyote problems adjoins a several-hundred-acre military preserve. And so on.

We have put out a welcome mat

We like our living areas to be attractive and as natural as possible, so homes, streets, roadsides, highway medians, and the like are full of bushes, trees, and flowers. This attracts rodents large and small, and that attracts coyotes. Many of these mini-greenbelts are large enough to support permanent colonies of coyotes.

We are feeding them

Feeding wildlife is such a satisfying and seemingly meritorious pastime that it is hard to get people to see how it could possibly be harmful. So people do it and keep on doing it, even when there are local ordinances against it. But feeding coyotes (or the establishment of food stations meant to feed other wildlife that coyotes can take advantage of) is the single biggest reason for problem coyotes, and coyote attacks—all of the experts are in agreement on this. It is bad enough that many urban coyotes have become accustomed to human scent and structures—how could they not, living right amongst us as they do? They have come to associate the plentiful food sources of suburbia—or state or national park campgrounds—with food. But when we go one step further and directly feed them, trouble quickly follows. While investigating coyote attacks, officials often have found that someone in the neighborhood was giving coyotes handouts. For example, coyotes were being enticed into viewing range by hamburger, pork chops, or even that hunk of venison someone gave you, which until now you never were able to figure out what to do with. Coyotes don't need our help to survive, and once coyotes associate humans directly with food, they are all the more likely to get ever bolder—start coming around more often, including in the daytime, help themselves to our pets, and start eyeing our children. See pages 98–99 for more about the disastrous habit of feeding coyotes.

We no longer seem to be a threat to them

Coyotes were pursued relentlessly through most of the past century and a half—with traps, guns, poisons, dogs, and more. In more recent years, especially in urban and suburban areas, there is no hunting, no trapping, no deadly poisoning campaigns. And much of the population is all in favor of wildlife of any kind. This has made coyotes, which are nothing if not smart, less afraid of people in urban and suburban areas. They not only live among the sights, scents, and structures of humans, they see no negative consequences now for living close to their old enemy.

So generation by generation, urban and suburban coyotes have lost some of their fear of man, and some have even become aggressive toward humans.

Plain old hunger

Some coyote attacks can be attributed to what the experts call food stress, or plain old hunger. This is most likely to happen when coyote parents are trying to keep up with the daily demands of a litter of pups, or when a male coyote is trying to help feed his very pregnant mate. At such times, the abundant resources of the suburban environment notwithstanding, coyotes will increase their preying on pets, and perhaps even threaten our own young ones.

Children look like prey

Young children, such as toddlers, especially, are in the size range of typical coyote prey, and the fact that they may crawl, be lying down, or otherwise not look like upright walking humans, only increases the resemblance. Young children, playing or otherwise, may also come out with shrieks and other high-pitched cries that resemble the ones coyotes have come to associate with an animal in some kind of distress that is easy prey. Most coyote attacks on children have been on those under ten years old.

Movement

Coyotes, like most other predators, have a built-in instinct to attack anything that moves quickly, or runs from them. This has worsened many of the attacks on adults as well as children.

Trying to protect pets

Many adults, and some children, too, have been bitten or scratched by coyotes in the wake of their efforts to keep a beloved pet from being killed or spirited away.

Can we do anything to reverse this trend?

We can, as long as we do it as soon as possible. The coyote-discouraging techniques described in the next chapter will help coyotes remember that humans are not to be messed with. But these things must be done consistently, by the entire neighborhood if feasible. And to really work they have to be done as early as possible, when coyotes first start getting bold (although it is never too late to stop directly feeding coyotes). As coyote aficionado and author Catherine Reid says, coyotes mark their own territory with urine, their scent glands, scat, and howls. You need to find a way to mark your own territory—your yard—"in a way they can understand. When they get to the edge of your property line, that's where you stop [your discouragement techniques]. They'll know that's your territory."

Have neighborhood meetings to make sure everyone, child and adult, knows what a coyote looks like and what they can do to help prevent coyote attacks on people or pets.

Once coyotes have firmly established the habits that lead to trouble, they must usually be removed by authorities and euthanized. As one wildlife officer explained, "This is not done to make the guilty party pay, but to stop the cycle of learning. If a coyote passes on that knowledge to its offspring, things become more dangerous."

What if a coyote actually attacks me?

If a coyote (or more than one) seems to be considering the idea of attacking you, practice some of the scare-off techniques outlined on page 111. Look the coyote directly in the eye, and keep eye contact with it. Make yourself look as big as possible, by standing tall, standing on something nearby, and fanning out a coat or jacket if you are wearing one to make yourself look larger. Retreat from the animal or animals by walking backward slowly, toward a building, vehicle, or group of other people—don't turn your back on it.

Should you actually be attacked, here are the things, first, NOT to do: Don't shriek or scream if you can help it, because it may reinforce the idea that you are prey. And don't run—fast movements like this trigger the attack response in many predators.

If you are charged by a coyote, and have anything in hand or nearby that could conceivably be used as a weapon, use it. If you are weaponless, kick or punch the animal hard in the nose if you can. Use your arms to protect your face and throat, and if you fall or are knocked down, curl up into a ball with your arms and hands over your head and neck.

If all else fails, try to grip an attacking animal by the throat—do whatever you can to keep its teeth off you. As coyote researcher Stuart Ellins has noted, "Although their jaws are not as strong as those of larger carnivores such as wolves, a coyote's mouth can be a lethal weapon." When Ellins was giving a captive coyote an injection once, "her flashing teeth sliced the flesh on my hand like razor blades, and she imbedded her daggerlike canines into my arm through heavy clothing with the precision of surgical instruments. Indeed, hers was a mouth created not only to tear meat but to kill."

Getting help

Call 911 immediately if you can, of course, and shout for the help of any other people nearby. Be sure to report any coyote aggression or attacks to not only your local law enforcement agencies, but the health department, and department of fish and game. If you have been bitten or scratched by a coyote, until medical help arrives, wash the areas involved well with soap and water, being careful to avoid getting any coyote saliva near broken skin, scratches or scrapes or the like, or on mucous membranes such as those in your mouth and nose. Rabies can be spread that way, even if you were not bitten.

Since rabies in humans is almost always fatal, you may need to undergo rabies immunization—without delay, if the doctors say this should be done—and authorities will attempt to catch the coyotes involved in the attack to test them for rabies.

Protecting children from coyotes

One never wants to leave young children unwatched in any case, but if coyotes have been getting bolder in your area, keeping a close eye on them at all times becomes even more important. Many coyote attacks on youngsters have occurred while a parent left for only a minute or two to do or get something, and some have happened in spite of the fact that the parents were right there. If you are in the yard with your child and see a coyote, pick the child up right away and take it inside. Be especially watchful when children are playing near a pool or other water source, with a pet, at a campsite of any kind, or near woods or roads or other corridors that coyotes use as travel routes. Remove brushy vegetation from and around children's play areas.

coyote control:
Ways of Discouraging Coyotes

The subject is almost as fertile a field for myth and legend as how to get rid of moles, but here you will learn how coyotes really feel about lights, noises, human scent, chemical repellents, and more. Difficult as it is to effectively discourage coyotes, this chapter will explain all of the ways that have real merit, while debunking those that don't. It describes all of the different things you can do to give your pets and your family the best chance of avoiding a dangerous brush with coyotes, as well as what you may unintentionally be doing to attract them.

All in all, you will find out here how to make your home grounds less hospitable for coyotes, without destroying your own pleasure in it.

HOW CAN I KEEP COYOTES OUT? CAN I FENCE THEM OUT?

A good title for this topic might be, "under, over, around, and through"!

Fencing coyotes out is, as they call very difficult endeavors, "a challenge." There are probably no truly coyote-proof fences, though there are a few pretty coyote-resistant ones. None of these are simple or inexpensive

Photo by D.J. Hannigan, law enforcement, Park County, Colorado

to construct, so they are usually employed where money and effort are no object to protect a small area of something quite valuable, or cherished. And even these coyote-resistant fences work best with a secondary line of defense as well, such as livestock guard dogs, as discussed in Chapter 4. When attempting to build a coyote-stopping fence, you may want to do a little cost–benefit ratio calculation first. You need to take into account, too, the zoning or other regulations that apply to fences where you live.

How do coyotes get through fences?

First let us consider how coyotes normally cross or get through fences. Unless a fence is short, they don't normally leap it in a single bound, like deer. Their first choice is normally to squeeze or dig under it. This is usually done at places where the fence wire doesn't quite meet the ground, where the fence crosses a small gully or washout, or at places like gates, where there are gaps near the ground.

Normally, their second choice is climbing a fence, which they do by gripping the top with their front paws and propelling themselves up and over with their back legs, as dogs do. This is why they are sometimes found with their feet caught between the top barbed wire and the very top of the mesh wire of a woven wire fence. Once they get a good start climbing, coyotes can scale five- or six-foot fences of any conventional kind, and even 8- or 12-foot-high cyclone fences. In climbing or jumping, when they do jump, they may use fence crossbraces and corners to give them a good start or foothold.

Electric fences

Wholly electric fences are usually used in rural areas to protect livestock. If properly built, taking advantage of the latest technology (much of it developed in sheep-minded New Zealand and Australia), such as low impedance chargers, they can be as much as 80 or 90% effective against coyotes. In general, electric fences are less expensive to build than wire fences, in terms of both materials and labor, but over the long run, they require much more maintenance.

A six-wire electric fence can keep coyotes out of an enclosed area. Drawing by Jenifer Rees and used by permission of Washington Department of Fish and Wildlife

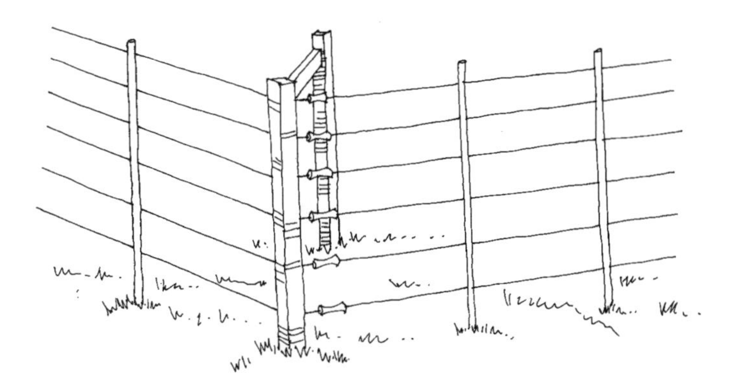

The most effective design for coyote deterrence is an electric fence that is four to four and a half feet high, composed of from 6 to 12 smooth high-tensile wires (7 or 9 is better than 6, and 12 of course is best). The wire closest to the ground should be only five inches from it, and no wires should be farther apart than ten inches. The fence needs to be well braced, with posts not too far apart, and the wire should be tightened to a tension of 200 to 300 pounds. Some such fences alternate charged and ground wires, though the latest designs have all charged wires. The wires should be charged to a minimum of 2,000 volts, up to 4,000. The most effective fences also have a charged wire just above the ground, eight inches high, outside the fence.

Fences can also be built of electrified netting with fiberglass posts built right into it, which is what *The Daily Coyote*'s Shreve Stockton chose to keep her pet coyote in, rather than out.

Keeping fences electrified

The necessary maintenance mentioned above boils down to all of the steps you may have to take to make sure that your electric fence stays "live." This includes:

- Keeping it clear of things that will ground it out, such as grass and other vegetation growing up/under/against it, branches falling on it, water pooling near it, or rodents burrowing under it. Herbicides can be used to control plants around a fence, with care.
- Repairing damage done to the fence by livestock and/or wildlife.
- Making sure the wires stay tight, and that the charge is sufficient.

This gives you a hint of what is involved here, and an excellent reference for electric fence builders is "21 Mistakes to Avoid with Electric Fencing" (**pasturemanagement.com/mistakes.htm**). And "Building an Electric Antipredator Fence," by Oregon State University, is a thoroughly detailed guide to the whole process (**extension.oregonstate.edu/catalog/html/pnw/pnw225/**).

Wire fences

A fence made of wire must be taller than an all-electric one. This means five and a half to six and a half feet tall, and made of woven or welded wire with small "stays," or openings. The openings in a fence meant to exclude coyotes should not be larger than four inches on a side—coyotes can squeeze through the six-by-six-inch openings of many ordinary field fences. The fence must be tightly flush to the ground, and the wire stretched tight, and solidly anchored. The fence wire should either be buried six to eight inches in the ground, or have an 18- to 24-inch apron of the same wire the fence is constructed of, which needs to be solidly attached to the bottom of the fence on the outside.

Fence extensions are needed to keep coyotes from jumping over a five-foot fence. Angle the top of a woven-wire fence out about 15 inches as shown. A fence intended to keep out coyotes also needs to be buried at least a foot in the ground or have a wire apron on it, as shown here. Drawing by Jenifer Rees and used by permission of Washington Department of Fish and Wildlife

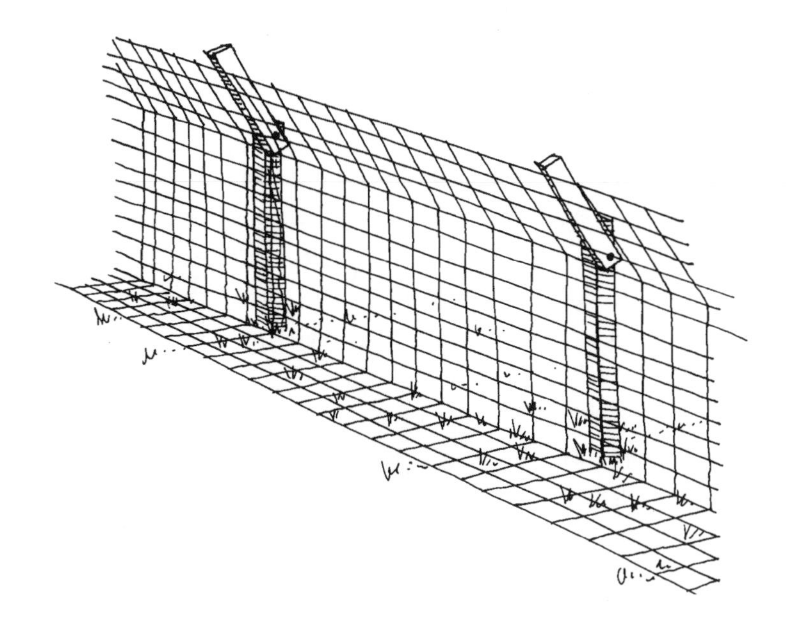

This alone is not enough, however, if we remember how coyotes get through fences. There also needs to be a wire overhang at the top of the fence, 15 to 24 inches wide, set at a 45-degree angle to the fence posts, on the outside of the fence. Some people add a strand of barbed wire or a hot wire at the end of the overhang. You have probably seen arrangements like this on the fences of the "big cat" enclosures at the zoo, and for the same reason—to prevent climbing—in this case, out rather than in.

Wire plus electric

If you have an existing wire fence, such as woven wire, in good condition, you can add some electrified wires to it for one of the most coyote-frustrating fences. Just add two or three (or more) high-tensile wires, one at the very top of the fence, and the others outside the fence, offset from it by eight to ten inches, on offset brackets. After the top one, the others can be at the middle and bottom (five or six inches above the ground), for instance.

Suburban fencing

Many a pet, cats and small dogs, have been snatched from or attacked in the ordinary fencing of suburban and city yards. So if you want to leave your pet outside with any real degree of safety, you need to consider one of the following.

- A totally enclosed area. This means an area that is fenced sides and top with sturdy wire. Since cats can squeeze their heads through spaces less than two inches wide, a good choice is strong, closely set wire, such as two-by-four-inch welded wire with an overlay of vinyl-coated chicken wire (ordinary chicken wire rusts out quickly). The entire enclosure needs to have sturdy supports of treated lumber and well-secured posts, preferably set into concrete. Remember what good diggers coyotes are, and either sink the wire into the ground six to eight inches, or add a sturdy 18- to 24-inch wire apron all around the bottom of the enclosure.

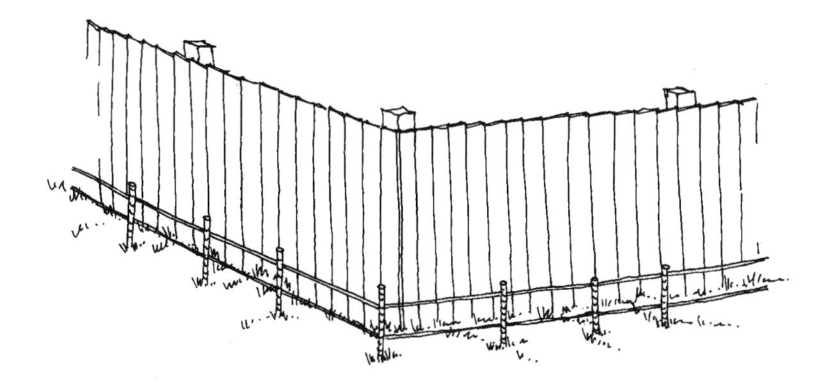

Two electrified wires, 8 and 15 inches above ground and offset 12 inches from an existing wooden fence, will keep coyotes from climbing the fence. Drawing by Jenifer Rees and used by permission of Washington Department of Fish and Wildlife

- If you have a tall, wooden fence of the privacy type, you can add two electric wires outside it, 8 and 15 inches above the ground, if electric fencing is allowed in your area.
- To help prevent coyotes from scaling a fence of any kind, you can add what is called a "coyote roller" all along the top (see **coyote roller.com**), or construct your own version of this device with PVC pipe supported by a tight, strong wire at the top of your fence. This keeps coyotes from getting the purchase they need to vault over.

CUTTING DOWN ON COYOTE CONFLICTS

As city coyotes grow bolder, are seen more, and attack more pets and people, the number of complaints about them climbs. As one suburbanite put it, "The coyotes are so brave now, you flip the light on and still have to go out and chase them away."

The best way to deal with coyote conflicts is to prevent them. Here are some good ways to do that.

COYOTE CONSCIOUSNESS FOR CAMPERS

In campgrounds, too, avoid feeding coyotes or leaving food where they can easily help themselves to it. If there have been coyote incidents in the area where you are camping or hiking, keep a coyote discourager handy (see page 110) or carry one with you. When sleeping outdoors, you are safer from coyote harassment in a closed tent than in a sleeping bag just stretched out on the ground.

Coyote-proofing a yard

Coyotes have no concept of property lines—your front yard and backyard are just pieces of the territory they traverse every night. Your goal is to make your property as unattractive or inaccessible to them as possible.

- This can start with **fencing** your property—not with a wimpy decorative fence of some kind, but one that gives you the best chance of keeping coyotes out, as described earlier in this chapter. Many of the pets snatched and children hassled by coyotes in urban areas have been in yards fenced the wrong way to protect them from this particular threat. Once you have a good fence, inspect it regularly to make sure there are no holes in it, or dug beneath it. And keep grass and shrubbery from growing up around it.
- Coyotes love **brush and dense vegetation** of all kinds—they sleep in it and hide in it—so your next step is to remove or trim back any thick greenery in the yard, or near it, that will provide them with cover. This means tall grass and weeds, overgrown shrubs and evergreens, and even dense foundation plantings. The branches on the bottom two feet or so of trees and shrubs should be removed to eliminate hiding places. If there is a field or vacant lot nearby, see that it is mowed or bushhogged. Not all of these measures will

please the most green-minded among us, but they will undoubtedly give coyotes fewer places to lurk undetected.

- While you are neatening up the yard, get rid of old **woodpiles** (which amount to rodent apartment houses, as well as posing a termite risk), **junk piles** (another mouse and rat haven), **brush piles, old logs,** and the like. If you have a **fountain** or **mini-pond** in your yard, and the yard is unfenced, consider getting rid of these coyote waterers, or fencing around them.
- **Close off crawlspaces** beneath decks, porches, mobile homes, and the like, because they provide handy lounging and denning places for coyotes.
- Keep the **barbecue** grill clean, to eliminate seductive odors, and, if possible, put it away when you're not using it.

Sound and lights!

You may also wish to install motion-sensitive lights, or floodlights, for nighttime. Coyotes are repelled to some extent by lighting. The flash of a game camera will also scare off many coyotes.

Recordings of the human voice (finally, a good use for talk shows!) can spook coyotes, and on the Internet you can even find recordings of cougar growls and screams. The cougar is a true top cat, which preys on coyotes, and the producer of the CD swears that the recording is a true coyote deterrent. I played one on my portable CD player, and it was impressive, although if you played one of these outside you would definitely need to warn the neighbors first. Big noisemakers like propane cannons can also be used to drive coyotes away, if you live where this is practical and legal.

Coyotes are smart and observant, so anything that makes a fuss or racket, no matter how impressive, will eventually lose its effectiveness after the coyotes have a chance to see that it happens over and over and no real harm to them follows.

The experts all say that the best coyote deterrent in the "sound and lights" category is one that combines more than one off-putter, and varies the way these things are served up. There are various devices on the market that do just this, including combinations of flashing lights and sirens such as the Critter Gitter, the Electronic Guard, and the Night Guard Solar. But even these, if they can be used where you live without enraging the neighbors, will work for a couple of months at most before they, too, are judged empty threats by the wily coyote.

Protecting children's play areas

Take extra care to reduce possible coyote hideouts in children's play areas. Prune off the bottom branches of trees and shrubs to give you full visibility. Never leave young children unattended for any length of time in a yard, even a fenced one. If there have been coyote sightings in the area, keep a coyote discourager—a broom, pile of stones, hockey stick, or old golf club—handy.

Garbage patrol

Although the urban environment offers coyotes plenty to eat besides the leavings we bundle into plastic bags at the curbside, coyotes (like raccoons, stray dogs and cats, and other creatures) will search for goodies in the garbage if we give them a chance. Coyote-proofing your trash will help keep other critters out of it, too.

First of all, use sturdy cans, metal if at all possible, with tight-fitting lids. Use straps or bungee cords to attach the lids firmly otherwise. If you cannot keep the cans in a garage, shed, or enclosure, consider attaching them to a post or the side of a building. If you wait until the very morning of garbage collection day to put your cans out, you stand much less chance of spending a half hour or so picking up a repulsive collection of rummaged-through discards strung out all over the lawn or roadside. For extra protection, some people spray ammonia or pepper spray into bags containing meat, cheese, or other attractive scraps, and clean their cans regularly with hot chlorine bleach solution to remove food odors.

Guarding the garden

As noted earlier, coyotes are very fond of fruits and vegetables, and they love berries, corn, melons, and even tomatoes, so gardens are not immune from attack. The answer here is good fencing, as described earlier, although some people use heavy tarps in a pinch to protect ripening vegetables and berries at night. Keep fallen fruit around fruit trees picked up, or fence orchards, and you might want to avoid landscape trees and bushes that produce edible fruits and seeds. Coyotes in some desert areas subsist for parts of the year almost entirely on things like mesquite beans, juniper berries, and cactus fruits.

Alternative ways to try and discourage coyotes from raiding your crops include putting bars of strongly scented deodorant soap on the boundaries of your garden, and spreading human hair (from your friendly local haircutter or salon) there. Wolf urine is also said to repel coyotes, but finding it, and affording it, might be a problem.

Compost piles can also attract coyotes and other prowlers. The best approach here, which is what all of the garden guides tell you anyway, is to never put meat scraps or other animal matter in a compost pile. To keep coyotes from foraging for worms and grubs in the pile, fence the pile, house it in a wooden or hardware-cloth–sided box with a top, or keep it securely covered with a sturdy tarp when not in use.

Bird feeder foresight

Bird feeding is more popular every year, so there are a lot of feeders out there. But if not properly managed, they, too, can precipitate coyote problems. Don't let fallen seed build up under feeders—choose hanging styles that minimize fallout and clean it up regularly when it happens. Fallen seed means rodents and birds on the ground—a real draw for Wily. Put feeders away at night, and remove them entirely for a while if you are having coyote conflicts.

Pet protection

Most of the pets-snatched-from-backyard horror stories started with coyotes being attracted to pet food. Those ingredient engineers at the pet food companies did a good job—pet food, dry or wet, is just as attractive to coyotes (and a wide range of other wild animals) as it is to your cat or dog. Feed and water pets inside if at all possible, and if you must feed outside remove any uneaten food, and water bowls, well before dark. For this same reason, never store pet food outside unless it is in sturdy and tightly sealed containers.

Don't leave food out for feral cats or dogs, or raccoons, or anything else. It will only draw coyotes, which may eat the animal you intended to feed as well as the food you left for them.

Don't leave dogs tied outside, even in a fenced yard. Coyotes are well aware that a chained or otherwise restrained dog is a pretty helpless one. Likewise, invisible fences do nothing in the case of a coyote invasion, except put a dog at a disadvantage. Outside dogs should be in a secure kennel at night—which means one with a top, not just high sides. Coyotes are good climbers. Keep any pet that weighs less than 50 pounds in at night. Cats are safe outside only in a sturdy roofed enclosure (see page 105).

For protection from coyotes, "hutch" pets like rabbits and guinea pigs need to be in a sturdy cage with a solid bottom—not a cage suspended up off the ground, with only a wire bottom, handy as that is for poop disposal. Give pets like this a little house within the cage, too, to hide in, so they don't have a nervous breakdown while a coyote is scoping the situation out. Hardware cloth is the ideal material for cage sides and tops, to keep predators from being able to reach or nose their way in.

Poultry needs to be in a secure, closed building at night—not just a fence of flimsy chicken wire. Any fence in a chicken yard needs to have sturdy posts and be buried a foot in the ground, or have a wire apron (see page 125) to keep coyotes from digging in. If coyotes (and aerial predators such as hawks) are a serious problem in your area, consider a well-supported wire top for the yard or pen as well.

It's a good idea to remove tall weeds and brush from the area around poultry or rabbit pens or yards, to make it harder for predators of all kinds to sneak up within striking range.

WHAT CAN I DO IF A COYOTE APPEARS IN MY BACKYARD?

If there is nothing the coyote could harm out there, and you just want to enjoy watching it, see Chapter 8.

If you want it gone as quickly as possible, do the following.

First of all, if you show yourself to the animal, you absolutely do *not* want to appear indifferent to it, to run from it, or try to pet it or otherwise be nice to it. The only way to keep urban coyotes in a position to stay alive, and coexist peacefully with people, is to keep them afraid of people.

Stand up if you are seated. Immediately pick up any children or small pets and move them inside, or shoo big pets in. If it's nighttime, turn on any yard lights, or grab a powerful hand lantern or big flashlight, if you have one, and shine it on the animal(s). That alone may make the coyote leave. If it doesn't, charge at it, yelling as loudly as you can. You can also stomp your feet or clap your hands. Other good ways to make noise include clanging pots and using an air horn, car horn, or starter pistol. You can make yourself a "coyote chaser" to keep handy for such moments by putting metal washers or pennies into an empty pop can, covering the can with aluminum foil, and sealing the foil well with duct tape. You can also fill a metal cookie tin about a quarter full of nails, or connect several empty cans together with string, which will make a good clatter when tossed on the ground. A "coyote clacker" can be assembled from two two-foot pieces of two-by-four wood hinged together at one end.

If noise and light don't send the coyotes away, you can throw rocks, sticks, or tennis balls, or squirt the invader with a stiff spray from the hose or a Super Soaker water pistol filled with water or vinegar. Moving up on the scale of aggression, if it becomes necessary, you can chase the coyote with a broom, shovel, golf club, pitchfork, or baseball bat, acting like you mean it, or fire a slingshot, pellet gun, air rifle, or shotgun loaded with rubber pellets at it. Some people have even resorted to paintball guns. Rural residents by this time might be getting the shotgun from the gun safe.

If none of the above does the trick, and the animal appears unafraid, ill, or confused, or barks or growls at you, get inside and call your local animal control officer, or Division of Wildlife, immediately.

In your efforts to evict it from your property, be careful not to corner a coyote that is actually trying his best to head for the hills.

EDUCATION AND COMMUNAL EFFORT: KEY TO COYOTE CONTROL

Keeping coyotes from becoming a problem in urban and suburban areas needs to be a neighborhood effort, not just an individual one. No matter how hard you try to assure the safety of your own children and pets, if someone else down the block is feeding coyotes on their deck every night, your efforts will get you nowhere.

One of the best approaches is to have a neighborhood meeting on the subject. Make sure that everyone, adult and child, knows what coyotes look like, and that they are wild animals that need to be left alone and never approached. Children especially need to understand that coyotes are not dogs or pets, and that they should tell you immediately if they see one. If one person's pet is attacked, all the neighbors need to know, right away, so they can take measures to protect theirs. And everyone needs to be 100% clear on the fact that feeding coyotes in any way is the surest path to grief and trouble.

Ideally, the experts point out, towns and cities that adjoin each other should also have uniform laws and ordinances, to make the enforcement of measures meant to help reduce coyote conflicts more manageable.

CHAPTER 7

when coyotes must go:

trapping and hunting coyotes

Trapping and hunting coyotes (or what is euphemistically called in coyote literature "lethal removal") is a challenging subject—trapping coyotes, for instance, has been called "the Super Bowl of trapping"—as well as one rife with myth and misinformation. This chapter presents a concise guide to the subject, for the sake of readers who have concluded that a gun or trap has become necessary, or is the only answer. Included are the best and most effective means and measures in the areas of both hunting and trapping: where to hunt or trap; the best tools for the purpose; baits; attractants; ways to minimize scaring away your quarry; what to do with a coyote if you manage to catch one; and more. Here you will find plenty of advice from those who successfully hunt or trap this most elusive of predators.

REMOVING THE ACTUAL OFFENDERS IS THE BEST APPROACH

As noted earlier, it is almost impossible to remove every single coyote from an area permanently. Even if you could manage to kill every coyote there today, before the week was out reinforcements from the endless pool of traveling coyotes called "solitaries" or "transients" would rush in

A western coyote on a mountainside in Yellowstone. Photo by John H. Williams

to fill the void. To actually have a significant effect on a coyote population, three quarters of the resident coyotes would have to be killed every year. If you did that, within a mere 50 years or so, you might actually have reduced or eliminated the population.

These realities, plus of course simple humaneness, mean that the best approach almost always is to remove only the coyotes actually causing problems or damage.

In rural areas you might want to take this on yourself, if you feel comfortable with such things, by the means outlined in this chapter. In urban areas, where many of the ways normally employed to kill coyotes are impossible or very difficult to put into effect, it has to be gone about differently.

Check the rules first

The first question always is: What are the rules regarding coyote removal in your area? In some states coyotes can be hunted or trapped year-round; in others this can be done only in certain seasons, and by specific means. In some locales landowners can go after a coyote inflicting damage on his livestock or pets without permission, in others a permit may be needed, or authorization given. In some states coyotes are varmints to be eliminated at will, by just about any means; in others they are game animals or furbearers that can only be pursued according to the state rulebook. Find out first, before going after Wily.

The best time to remove problem coyotes is usually after the pups of the year have dispersed, and before the birth of the new crop (see Chapter 3), since adult alpha pairs with young are usually the greatest offenders.

Help with coyote removal

In urban areas, especially, coyote removal is best attempted by people with real experience in the undertaking, or professionals in animal removal. Even local law enforcement agencies (often called upon to deal with coyote emergencies) turn to the specialists, because inept attempts to remove troublesome coyotes will only worsen the problem. Your county cooperative extension agent, the Animal Damage Control Unit of the U.S. Department of Agriculture, or state wildlife agency can help you here. Some of these sources can provide experts to help you assess the situation to be sure a coyote was the culprit, and if so, find an answer for your particular problem. Experts like these not only have a better chance of capturing Wily, they can use means that may not otherwise be legal in your area, and use them correctly. Another resource here is commercial businesses that specialize in helping urban householders deal with nuisance wildlife. In urban areas removal experts may use sharpshooters poised over bait of some kind, or officers armed with shotguns or small caliber rifles, sometimes with silencers, used only when a clear shot is possible. Night hunting with lights is sometimes employed, near well-established coyote trails. Where legal, padded leghold traps or snares may be used. In rural areas, aerial gunning and M-44 devices are sometimes used.

JUST HOW HARD IS IT TO TRAP A COYOTE? IS THERE ANY HUMANE WAY TO DO IT?

The trapping of animals is one of the oldest human undertakings. As far back as neolithic times, and probably before, people were scheming up ways to catch unwary animals for the cookpot. Before Europeans arrived in this country, Native Americans used a number of different kinds of traps to help feed themselves.

In more recent times in the United States, trapping often has been a way of appropriating the handsome coats of animals for our own purposes, or a means of attempting to stop their depredations, real or imagined. Top-quality coyote pelts fetched as much as $150 back in the 1980s, before fur prices of most kinds dropped precipitously in the wake of the anti-fur movement. Today, most of the coyotes trapped are part of a removal effort, or taken by accident by people trapping foxes.

The handsome winter coat of the coyote. Photo by John H. Williams

The Super Bowl of trapping

Many coyotes have been trapped since the days of the French fur trappers, but it has never been easy—in fact, it has been called "the Super Bowl of trapping." Given the coyote's considerable intelligence, tremendous aptitude for learning, keen senses, and native wariness, it is easy to see why this would be so. Coyote literature is full of stories about how masterfully coyotes have evaded traps of all types, and of the ways they have expressed what seemed to be their contempt for the process: by digging out all around a trap, for example, and then urinating on it. Successful trapping of any animal is an art, usually learned by hands-on experience, and in the case of coyotes, it is an art you had better learn quickly, or you are just worsening the problem. Inept trapping efforts will only make coyotes far more difficult, if not impossible, to catch in the future.

Young males are easier to catch than females, except for mothers out hunting to feed a brood of pups.

The most effective methods

To complicate things further, the methods with the best chance of success for coyote removal are highly controversial ones at this point in time. The leghold trap and the snare give you the best chance of actually catching a wily coyote, but even if these devices are legal in your state, and you are willing to use them, there will be plenty of people who wish that you did not.

Leghold traps have been banned by eight states and at least 88 countries for the following reasons. These traps often injure an animal, which can mean anything from broken toes to compound fractures of the leg to exposing the animal to frostbite, and can lead to the famous self-mutilation of an animal chewing off its own foot or leg to escape. Once caught, the animal has to wait in pain and fear for the trapper to come and dispatch it. If the trapper does not appear as quickly as he or she should, the caught animal may be torn to pieces by other wildlife (including coyotes), or suffer hunger and thirst. And then there is the far from minor issue of what is euphemistically called the "nontarget species" often caught in such traps.

But if you really want or need to be rid of a problem coyote, the leghold trap, if legal in your area, is still probably the surest way to catch one. (Not that it is easy! See the following.) An effort has been made in recent years to make traps of this type more humane, with things like rubber padding in the jaws, offset jaws, and even tabs of quick-acting poison or tranquilizers to lessen the mental suffering. Researchers and conservationists often use padded leghold traps, such as the Soft Catch, where they are legal, to trap animals for study when necessary.

As nasty as leghold or foothold traps may be, you might be willing to use one if you have suffered serious predation from coyotes on livestock, or your pets have been disappearing into the gullets of coyotes, never to return. In such a case, you can make yourself feel somewhat better by considering the fact that whenever you catch a coyote, since relocation is frowned upon everywhere, you will have to kill it anyway. And as one longtime animal removal expert pointed out: "My job is to catch and kill animals. No matter how I do it, how humane is death? And for that matter, nothing is crueler than nature. Have you ever seen a fox caught in a trap, half-eaten by coyotes?"

The how-to of trapping

There is a lot to using a steel trap for coyotes effectively, which is very clearly outlined, described, and illustrated on the following sites, among others (**www.iafwa.org** and **icwdm.org/handbook/carnivor/coyotes.asp**), and in Texas Wildlife Damage Management Service publication L-1908 and publication For-37 ("Managing Coyote Problems in Kentucky"), from the University of Kentucky.

In general, there are a number of different ways a leghold trap can be set up, including the old traditional dirt hole set, the scent post set, the trail set, carrion sets, and blind sets. The key to success with any of these involves using the right size trap, picking exactly the right place to set it, minimizing human odor and signs of human disturbance you impart to the trap and the area around it, concealing the trap well, and choosing the right scent or lure to put near it. This is important, because steel traps for coyotes are not baited directly with food, but set close to something the coyote is impelled to investigate, such as a hole with a strong-smelling lure down in it, or a scent post within his territory marked with alien urine by you. The most common lure for coyotes is coyote urine (available in some hunting supply stores and by mail order), but many experienced trappers have their own favorite recipes and formulas, which include everything imaginable, from coyote anal gland secretions, brains, or footpads, to chunks of aged meat or fish, seal or fish oil, rotten eggs, rotting ground squirrel or armadillo, skunk musk, and port wine. Some trappers include curiosity-provoking features like a ticking alarm clock down in the hole, in a dirt hole set, or deer ears buried so that the tips stick up from the ground. Others always wait a day or two after setting a trap before adding an attractant, until their own scent has dissipated some.

The Best Places to Set a Coyote Trap (Coyotes' Favorite Travel Routes)

- Coyotes love trails of all kinds and old roads, so this is one of the best places, but the trap should be set slightly to one side of it, not right on it, and upwind of the trail
- On hilltops, ridges, or saddles of land
- Near prominent features that stand out in the landscape, such as isolated big round bales of hay, boulders, or small clumps of trees
- By waterholes or stream crossings
- In dry streambeds, washes, or canyons
- Along fence lines or near fence corners
- Near field gates and other gates
- By the edge of a woods
- Near where livestock have been killed
- Not far from carcasses of livestock or deer, but not right next to them (or you will catch buzzards and other smaller scavengers)

Snares and like devices

Snares are another long-established way of catching animals—when an animal happens into a loop set out, usually on a game trail, the loop tightens around its neck until it suffocates. Snares are simple, effective, and inexpensive, and usually made of wire today, but from this description you can see why they have about as many opponents as leghold traps. Still, they are a good way of catching coyotes, especially when wet or snowy weather makes the use of steel traps difficult, or when coyotes are constantly slipping under a pasture fence to harass livestock. Snares work very well in such a situation. The same sources mentioned earlier will explain how to construct and use a snare, should you need to use one. But there are newer variants of the snare that are more humane, though they work about as well, or sometimes better.

One of these is called a cable restraint, and in many places you must take a training course to use it legally. It is essentially a non-locking snare, and it can also be arranged so that it lets the animal loose if, for example, a deer is accidentally captured, or more than a certain amount of pressure is exerted on it. The Collarum is another such device, designed specifically for canines. When a canid unleashes it by pulling on a scented trigger, it springs a noose of wire up over the animal's head. If set properly it will then hold the animal safely, usually without harming him, as if on a leash, until the trap tender arrives. This makes it usable even in urban areas full of pet dogs, and it is in fact often used to trap stray or feral dogs.

Cage traps! Wouldn't this be a better way to go about it?

Certainly the idea of capturing coyotes in a Havahart or similar trap is the most appealing emotionally. Animals caught in traps of this type rarely suffer anything beyond chipped teeth, scraped noses, or broken toenails, as long as they are removed from the trap in a timely fashion. (Although this begs the question of what you do with the coyote after capture, if releasing it is not an option.) However, not only is a trap of this type large enough to catch a coyote huge, hard to move around, and expensive, it is less likely to actually catch one than the traps described earlier. Why would we ever think that the skittish, super-cautious coyote would be eager to enter an obviously man-made contraption that would place him in close quarters with no quick and easy way out?

Coyotes can be caught in cage traps, but not easily. The success rate is somewhat better with young, inexperienced coyotes, extra-hungry nursing mothers, and in urban areas where coyotes are more accustomed to the trappings of humanity. Since leghold traps are illegal in Massachusetts, coyote researcher Jonathan Way used cage traps to capture most of the coyotes in his studies of the urban coyote. One of his secrets of success was filling the traps with bait (roadkill or supermarket meat scraps), and then wiring them open so that they couldn't be sprung until the wily coyotes had *months* of exposure to the idea that they could go in there, get some quick eats, and get right back out safely. After catching a number of hawks he never intended to catch, he also began wiring the traps open during the day, even once he started actual trapping. Then he checked every trap not just once, but often twice a day.

Rob Drimones, a wildlife control agent in Florida, also cage-traps successfully, and he has the following advice. "Traps for coyotes should be used only for coyotes—having the scents of other animals, such as wild hogs, on there will be off-putting to them. We have the best luck with live bait such as a chicken, rabbit, or rat, which is put in one of the special small cages sold by the companies that produce large live traps just for this purpose. The bait cage is set inside the trap, so that it attracts coyotes without them actually being able to get to the bait animal. We set the trap out with bait, wired open, for several days before actually setting the trap. When choosing where to place the trap, we usually ask all the neighbors exactly where the coyotes have been seen. One of the best places for a trap is in a woodlot, where they sleep and den. We camouflage the trap some with pieces of the nearby vegetation."

Nets

One other more humane method of capturing coyotes worth mentioning is net launching, though it is not something the average private citizen would undertake. This involves setting out bait and then using small cannons, or a helicopter, to propel a large net into the air and hopefully over the heads and backs of a couple of coyotes. This is expensive, and takes a lot of practice—and some help—to pull off.

Rent-a-trapper

Since skill and experience really count when it comes to trapping, and specialized equipment is needed, many people, especially in urban areas, would rather hire someone else to do the job. There are companies and individuals that specialize in this, and often divisions of wildlife, animal control agencies, or local law enforcement officers have lists of licensed trappers in your area. Peace of mind does not come cheap, however. Some companies charge $50 to $75 per animal captured, or rates like $750 a week. Why a week, you wonder? Because capturing coyotes is not a quick and easy job!

IF I CAN MANAGE TO CATCH THAT COYOTE, I WON'T KILL IT. I'LL JUST TAKE IT SOMEWHERE AND RELEASE IT.

This idea is in about the same league as dropping off no-longer-wanted pets in the country "so they can find a home." The intention may be good—we want to be rid of something, and we don't want to kill it—but this is dodging the issue, and no favor to the animal in the end, assuming it is even legal to transport or relocate wildlife where you live.

First of all, the reason we want to get rid of a coyote is usually because it is causing some problem. Assuming we can successfully trap it, or have it trapped, if we cart it somewhere else, it will just resume its

A nine-year-old female eastern coyote from New York.
Photo by Robert Watroba

problem behavior somewhere else—that is, eat someone else's pet or raid someone else's garden.

Relocating problem animals, be they coyotes or coons or whatever, also creates a problem for the resident wildlife wherever we relocate them. Almost never is virgin territory available with no members of that particular species already occupying it. So the "dumped" animal ends up having to duke it out with the resident coyotes or coons for food, water, shelter, and cover. It can sometimes die of sheer stress after being dropped into unfamiliar territory. Wildlife diseases can be spread this way, too.

In the case of coyotes, perhaps the biggest reason not to try the relocation approach is that unless you take them far, far, away, they probably will end up right back where they started. Coyote researchers in Chicago, for example, transported trapped city coyotes to the country, and every one of them, within 48 hours or less, hotfooted it back toward the Windy City. They didn't necessarily make it, though—most were killed by hunters or traffic on the way. Coyotes born and raised in urban environments actually prefer that environment, and rural coyotes have territories that they remember and want to reoccupy. The comforting thought of "relocation" for problem coyotes is mostly wishful thinking.

WHAT ABOUT BOUNTIES? CAN'T THEY HELP REDUCE THE COYOTE POPULATION?

The idea of a bounty is often a compelling one. When some big coyote problem rears its head, or some coyote outrage has just occurred (such as the killing of folksinger Taylor Mitchell in 2009), there is a natural impulse to say or think: if people were paid for every coyote they killed, the population would be reduced quickly. And since the value of furs of all kinds, including coyote pelts, is far less now than it was in earlier years, bounties have seemed at times to be the only real incentive that could be offered.

But the fact is that after three centuries of trying this idea out, bounties have never managed to significantly reduce coyote populations anywhere in North America. Every state that ever offered a bounty now has more coyotes than it ever had before. There are a number of reasons for this.

First is the fact that coyotes are in general incredibly difficult to reduce or eliminate. As noted earlier, if *three quarters* of the entire population in an area were removed *every year,* after 50 years or so the coyotes there might be significantly diminished, or begin to disappear. The number of animals taken in the average bounty-hunting campaign comes nowhere near this percentage. And as noted before, coyotes have a built-in ability to replenish their numbers. When coyotes in an area are seriously persecuted, they have bigger litters and start bearing young in their first year.

It doesn't help, either, that the wrong coyotes are the easiest ones to kill. Inexperienced, exuberant pups and teenage coyotes are the first ones to be caught in traps or come to a hunter's call, but it is the seasoned mature adults who are the most likely to be responsible for the things that get people up in arms, such as livestock predation. The highly experienced, cagey coyotes—the troublemakers—take more time and skill to capture than the average bounty hunter wants to invest.

Then, too, from the bounty hunter's point of view, the more coyotes around the better—the more potential payoffs. Bounty hunters of the past are well known for things like releasing all females so they can go off and breed more coyotes. This has gone so far at times that people have caged and raised bountied animals, including coyotes, for a handy homegrown cash crop.

Out-and-out bounty frauds are not unknown, either. Bounty programs usually require some form of proof that the target animal was killed, such as, in the case of coyotes, a "scalp" (usually the ears plus a connecting strip of skin). Few county or state officials are experienced enough to be absolutely sure that they are not looking, for example, at fox or dog ears. Another time-honored practice is transporting proofs-of-capture from where they were "harvested" to the state that pays the highest for them.

A better idea

A better idea than bounties, most states have concluded, would be to take those taxpayer dollars and invest them in hiring skilled professional trappers or hunters to go after only the problem coyotes, when necessary.

CALLING ALL COYOTES: CAN I REALLY CALL A COYOTE RIGHT TO ME?

Calling coyotes is such a big part of hunting them that the two subjects are almost synonymous. But since people call coyotes to photograph them as well, we will consider the subject separately.

The reason calling is so important to coyote hunting, and photography, too, is that the elusive and intelligent coyote is so hard to locate in the wild that the best chance most people have of seeing one is to make it come to them. This is done by positioning yourself carefully and then making an assortment of noises that is supposed to lure, intrigue, or deceive a coyote into coming into gun or camera range. If done successfully, it adds a thrill to the whole process because the coyote is then in fact hunting you.

There are two basic kinds of coyote calls, the fairly small ones that are operated by mouth and hand, and the larger, more high-tech electronic ones. Each has its advantages. The mouth calls are much cheaper and lighter, and can make quite an array of different sounds, mostly meant to be prey sounds, once you get the knack of it. It does take some serious practice to make the right sounds come out, and your loved ones and neighbors may suffer some until you do. The experts favor open-reed mouth calls over closed reed, because of their greater versatility, and the fact that they are less likely to become temporarily unusable once they have been dampened by your breath in very cold weather. Another popular type of manual call is a little rubber bulb called a mouse squeaker, and there are also mouth calls designed to produce noises made by coyotes themselves, called howlers.

With the electronic calls, which play prerecorded sounds from actual birds and animals, you can make realistic noises right out of the chute.

The most advanced and glamorous of these not only have a giant library of available sounds, but even have separate speakers operated by a remote, so that sounds aren't coming from exactly where you are sitting—an advantage when a coyote is coming in slowly, scanning for possible dangers to him every step of the way. They are also louder, or can be made louder when needed, than the mouth-operated ones. The drawbacks of the electronics are higher cost (from $30 all the way up to $800), greater bulk and weight, and the fact that they use batteries, which as we all know, can run down at inopportune moments. The electronic calls also may have a great variety of sounds, but they always come out the same way, whereas on a mouth call you can vary calls in as many different ways as the situation seems to warrant.

SOS sounds

The most time-honored and popular calls made for coyotes, and other predators, are the so-called "distress" calls, which imitate an animal or bird in trouble. Creatures in trouble are distracted, and very likely disabled, so the ever-opportunistic coyote sees this as a chance for an easy dinner. The great classic of distress calls is of a rabbit in distress—the normally mute rabbit, cottontail or jackrabbit, can make some truly dramatic sounds when it is upset or endangered. Distress recordings are available of everything from kittens and dog pups to fawns and foxes, from crows and woodpeckers to coyote pups. When you are trying to imitate such sounds on a mouth call, as *Outdoor Life,* puts it, you want to sound like something that is being "violently killed, torn limb from limb—use your emotions to create a soundscape of terror." Not all distress calls need to be quite that dramatic, but they should at least sound like the creature in question is hurt, or being threatened.

The how-to of howling

Either mouth or electronic calls are also available to imitate the sounds of coyotes themselves, and when you are using a mouth call, this can be harder than sounding like a suffering prey animal. Coyote language is used most often to call coyotes during breeding season, in late spring and summer when half-grown pups are wandering around and often beckoned by their parents' calls, and when a hunter is trying to see how many coyotes are around an area, and where, by persuading them to answer his howls. By turning a dial or pushing the right button on an electronic caller, or following the instructions that come with mouth-operated howlers, you can make anything from the coyote greeting howl, to the "this is my territory" warning, to a "here I am—where are you?" inquiry, to the seductive sounds of a female coyote in heat, or the mournful sound of a lost pup. Howls of various kinds will often work better on coyotes that have heard so many distress calls from would-be hunters that they have become immune to them.

Sirens, either handheld or vehicle mounted, can also be used for locating, or taking a quick informal census of, coyotes. If using a siren is legal in the area in question, you can travel around to likely coyote bedding spots with one and sound it. The best time to do this is about a half hour before dark. If a coyote howls, yips, or barks back, note that place on a map or in a notebook.

The best time to call

For coyotes, the best time to call is a little different from your phone plan's best calling times. In general, it is the first two hours or so after dawn, and before dark. These are the times when coyotes are very alert and stirring around, beginning their hunts for the day or wrapping them up. During a full moon, just ahead of or after a change in weather fronts, and in very cold weather (when coyotes are very hungry) are good times, too.

Find a good spot to call from

In general this means a place not only likely to harbor coyotes, but where you are going to have a good chance of seeing them, if they respond to your calls, before they see you. This ideally is a high point of some kind, where you can see several hundred yards in front of you and to the sides.

Since coyotes love trails, you especially want a view of any trails headed in your direction, above all the downwind ones. When coyotes hear a call, they almost always come in downwind of the sound, to see if their eyes and nose confirm what their ears are telling them.

You want the sun at your back if possible, and to always be calling into the wind. Given the coyote's keen sense of smell, if the wind isn't in your face when calling, you're wasting your time. Sitting near some object, such as a fence, big round hay bale, large rock, or the trunk of a tree, that will break up your outline is important, too.

The spot you pick should also be near coyote attractants such as edges of any kind (including fencerows and the edges of fields or woods), gullies, water sources, and marshy areas. Areas rich in prey like rabbits, mice, or prairie dogs are good, too, as are areas near calving grounds of any kind. Game or logging trails and old roads of any kind also are favored by coyotes, as is the area around a suspected coyote den or anyplace you have seen coyote tracks or other signs. On the outskirts of towns, look for natural areas a couple of miles from where development begins, small wooded or brushy areas, dumps, or small ponds and their surroundings.

Make yourself invisible

Make yourself invisible, or as close to it as you can manage. This starts with wearing camouflage clothing of some kind. It doesn't have to be the fanciest matching suit in the latest Cabela's catalog, but it should be something that makes you blend into the surroundings of your calling spots. In winter this may mean wearing predominantly white (which can just be an old white sheet draped over your shoulders), in dark forested areas dark green or gray, light green in spring or summer, tan in desert areas, and so on. You can test your proposed camouflage by laying it out on the ground in the area you have in mind, standing off from it a ways, and seeing if it stands out or fits in.

To complete your concealment, put grease paint on your face and hands, and if you wear glasses, wear a hat that will shield your face from

A ghilli suit is one approach to total-body camouflage.
Photo courtesy of Burris and Waddell

the sun and help prevent reflections from your lenses. Some people also take measures to prevent reflections from rifles and scopes.

Making yourself less likely to be detected by coyotes means doing all you can to keep from broadcasting human scents, too. This means taking all of the steps outlined in hunting guidebooks to see that your clothes are as scent-free as possible, avoiding the use of strong-smelling soaps, fragrances, and even toothpastes, and using a cover scent, or scent eliminator, such as Hunter's Specialties Scent-A-Way, when you actually go out to call.

You also want to keep a low profile on your way in. Your calling expedition actually starts before you arrive at "the spot." You want to park as far as possible from the spot, get out of your vehicle quietly, and walk quietly in. When you get there, sit down—don't stand there, or walk around glancing in all directions. Scan the area discreetly with binoculars after you're seated, and above all (especially if this is your first attempt at calling) have faith and confidence in what you are about to do.

Put the call to your lips

Or your finger on the button, as the case may be, and start sending those sound bytes out into the world. Most experts agree that you should start out softly, in case there are coyotes hovering somewhere right nearby. The recipes for calling vary, but in general it is something like: Call three or four times, or for a minute or so, then wait for five minutes or more (longer if you are in the heavy cover of eastern coyote country). Repeat this pattern a few times, increasing the volume if needed. By the time you have been there for 20 minutes or a half hour, a coyote should have appeared, if one is going to. Most experienced callers like to move to a new spot after 20 or 30 minutes, but if you have a spot you are certain will eventually yield a coyote (such as right beside a heavily traveled coyote trail) you might want to stay longer. Your next calling spot should not be closer than 200 yards from your first one.

After you call

Watch all of the trails in sight, and any well-shielded routes coyotes are likely to take to get to your position. Crows, magpies, or scolding smaller birds can be a clue to a coyote on the move, as can spooked deer, or chirping red squirrels or ground squirrels. If you actually see the gray and tan ghost coming in, yellow eyes ablaze with anticipation, try not to succumb to coyote fever. Keep your brain in working order and gently pick up your camera or rifle and get ready.

How often can you expect to succeed?

Certainly not every time you stake out a place to call from. The experts say that if you actually call in a coyote about 40% of the time, you're doing fine. Better than that is phenomenal.

Calling no-no's

- Don't try to call if the wind is stronger than 15 miles per hour. Sound won't carry well and the coyotes may be edgy.

- Don't call without concealing yourself. You will just be creating coyotes that will never have any faith in a call sound again.
- Don't call too much or too loudly—a little silence and waiting can do wonders.
- Don't talk or move around while in your calling spot. This is why it's good to bring a little something to sit on, such as a small cushion, and dress in a way that won't leave you too cold or too hot. Anything that helps you stay still is worth doing or having.

HOW ABOUT HUNTING COYOTES? HOW HARD IS THAT?

It's probably the ultimate test of hunting ability, says Les Johnson, host of the TV show *Predator Quest*. Other experts agree that while brown bears, Dall sheep, and mountain goats are considered truly tough quarry, this continent doesn't offer a greater hunting challenge than the coyote. "It's like hunting the bionic man," said one would-be coyote hunter in New Jersey: "Their senses are so good, it's only a matter of seconds before they know you're there." No wonder that although 2,000 residents of the Garden State bought coyote hunting permits from 2000 through 2003, less than two dozen coyotes were actually killed by them!

Another testament to the difficulty of this particular undertaking is the fact that despite all of the macho types and superwomen running around doing their best to bring down a coyote, the majority of coyotes killed by hunters are accidents—coyotes that happened into the path of a hunter in a tree stand waiting for something else, or into the line of sight of a rancher on his daily or weekly patrol of his acreage.

The reason coyote hunting is so hard goes back to those exceptional coyote senses of smell, sight, and hearing, plus their innate wariness and seemingly limitless learning ability. On the other hand, its very difficulty makes coyote hunting attractive: "matching wits with the perfect predator," as one hunting magazine put it. "If I was closing in on a six by six bull elk . . . and saw a coyote, I would have to go after the coyote. I get such a rush out of fooling another hunter," admits Mr. Johnson. Adding to the fast-growing popularity of coyote hunting is the fact that you can hunt coyotes year-round in many states, and there is no bag limit in many cases, nor often much limitation on the means or type of hunting for them that is legal.

Whether or not one is personally in favor of hunting, the hunting of coyotes does have some genuine advantages. It helps keep coyotes afraid of humans—critical to our peaceful coexistence with them—and is in fact one of the most effective and cost-effective ways of removing problem coyotes.

Secrets of successful coyote hunting

CHECK WITH THOSE WHO KNOW Consult the people who are always driving around—mail people, package deliverers, rural utilities repair people, school bus drivers, police officers, game commission people, farmers, landowners, and deer and turkey hunters. Ask them where they've seen coyotes and when. This is often a clue to good potential hunting spots.

SCENT-PROOFING AND WIND DIRECTION The two most important secrets of success are a little abstract for many of us, so they are easy to ignore. But they really matter when it comes to the chances of actually bagging a coyote. The first is taking measures to scent-proof yourself, as described earlier. Since our own sense of smell is not sensational, it may be hard to believe this really matters, but it does. The second is paying attention to wind direction. Since coyotes can smell human scent 200 to 300 yards away or farther, it is critical to always pursue coyotes with the wind in your face, so it cannot carry your scent to them. The best coyote hunters always check the prevailing wind direction in the places they are planning to hunt, and check the weather forecast before they leave the house on hunting day.

SCOUT OUT AHEAD The most successful coyote hunters use a siren or howler and scan for coyote-rich areas, as described in the calling section above, to carefully scout before they ever hunt. When they locate promising places to call and hunt from, they note them on a map or draw a crude sketch of them, including that all-important wind direction.

GET PHYSICALLY FIT At least to the point that you can move around out there without strain or difficulty. A few months of sustained exercise (such as brisk walking, which doctors are always urging us to do) for 30 minutes every day can work wonders.

TIME OF YEAR The best times of year to hunt coyotes, according to some, are November and December, when many hunters are wishing for additional opportunities before year-end. Others say the best months are January (when coyotes are really hungry) and August (when inexperienced pups are afield, and easy to lure in or fool).

Firepower

What's the best gun for hunting coyotes? This answer could fill many pages in this book, if we were to explore every possibility, and every latest high-tech option. But the general consensus is when calling coyotes in, it's not a bad idea to have both a long-range and short-range weapon at hand. For long range, this means a centerfire rifle such as a 22-50 or .243, a .204, .220, or .223, with a scope. Bolt-action models are considered the most accurate. Newer options are guns like the DPMS L243 AR–styled rifle, which gives you the ability to make quick follow-up shots.

For closer in, at 40 or 50 yards, especially in wooded or brushy areas, you want a 12-gauge shotgun, preferably a semiautomatic and a three-inch magnum, with an extra-full choke to assure a tight pattern. The shotgun should be loaded with #4 buckshot or one of the fancier loads such as "Dead Coyote."

Some people use heavier guns such as the 30-06, 270, or Remington 375 magnum, but these are really overkill for the purpose. As one hunter put it, humorously, "One shot from one of these can take out a coyote and five or six more, if they are lined up right behind him."

No matter what kind of gun you use, support devices like shooting sticks and bipods will assure a much more accurate shot than shooting freehand.

If doing as little damage as possible to the pelt is an issue, very fast, fragmenting ammunition is available for some guns—check with your local sporting goods store or gun enthusiast.

Find a partner

When it comes to coyote hunting, two heads (pairs of eyes and sets of ears) are often better than one. Since most coyote hunting is done by calling, one can call and the other can stand by to shoot. The best place to put your partner is on the downwind side, since coyotes usually try to circle the caller downwind before they come in. Make sure you know where your partner is at all times, and agree in advance as to who shoots where. (More than two people in the field together, on the other hand, can complicate things and scare off Mr. or Ms. Coyote.)

Get high

As noted in the calling section, finding a high spot to call from is important. When it comes to hunting, this can also be done by hunting from a tree stand, or even setting up a camouflaged wooden stepladder and calling from the top.

Use a decoy

Many coyote hunters find that, especially in brushy country, adding a decoy to their calls makes all the difference. As master coyote hunter Tom Bechdel of Pennsylvania says, "For me hunting without a decoy is like hunting without calls—decoys have increased my success 800%." The purpose of a decoy is to give the approaching coyote something besides the hidden hunter to focus on—something it thinks is making those noises it has been hearing. All kinds of synthetic critters are available from hunting supply catalogs for this purpose, from plastic or fiberglass fawns, coyotes, and foxes, to battery-operated fake-fur covered things that vibrate and quiver. But Mr. Bechdel insists that the best coyote decoy of all is also the cheapest—one of those bedraggled leftover stuffed toy Easter bunnies you can find in the bargain bin after the holiday (minus the fake carrot or little Easter bonnet, of course).

Ready and set now

Be ready ahead of time. Before you start calling, be fully prepared to shoot. Make sure your rifle or shotgun is ready for action in every way. If you are using shooting sticks, test their height.

Estimate the distances to nearby landmarks so you don't over- or underestimate range. Don't shoot when a coyote is too far away for an effective shot—wait until he comes closer, or hold your fire. For a quick kill, the place to aim on a coyote is the head, or just behind the shoulders, where the heart and lungs are located. More coyotes, believe it or not, are missed at close range than farther away, so a prospective hunter should take the time to practice at all distances.

What about night hunting for coyotes?

If legal in your area, night hunting can be effective; when problem coyotes need to be removed, the authorities often hunt them at night with lights. Many hunters feel that the chance of a predator coming to a call is greater at night (coyotes do often feel more secure at night), and watching for eyes shining in the darkness as a coyote comes in makes night hunting an exciting pastime for many. Handheld, head-mounted, gun-mounted, and even vehicle-mounted lights are available for night hunting, complete with colored lens covers (red is the favored color) to make the light less visible to animals. Here is another good place for the buddy system, with one person calling while the other scans for glowing eyeballs with a light. Needless to say, knowing your partner's location at all times, and always being doubly sure as to what you are shooting at, is critical in night hunting. To give yourself that important extra measure of safety in target identification, you will need to call coyotes in closer when night hunting, than in the day.

I've heard that baiting can be a good way to draw in coyotes. Is that true?

If allowed in your state, baiting can be used to lure coyotes into view. This calls for either a good-sized dead animal (such as a road-killed deer, or dead cow) or a big pile of other bait, such as meat scraps and trimmings scrounged from local butchers or supermarkets before they ship them off to the renderer. Nuisance animal expert John Trout Jr. suggests laying out up to 25 pounds or so of such goodies (in a place that is generally attractive to coyotes, such as not far from cover), and then putting a layer of small logs over the pile, so that smaller scavengers can't snarf it all up before the coyotes get there. He also recommends wearing rubber boots and rubber gloves when handling bait, and checking the pile at the times

A FEW WORDS ABOUT BOW HUNTING COYOTES . . . AND MORE

"I often encountered coyotes while I was hunting something else, so I started carrying a mouse squeaker call and a fawn distress call with my deer hunting gear. When the deer hunting is slow, I take out my coyote calls, start using them, and then see what happens. Last year, for instance, a pack of seven coyotes came in to my calls one morning, and I was able to bag one of them.

When hunting coyotes with a bow I use a Hoyt Alpha Max 32 and Axis 400 arrows, with a Rage broadhead. There is a very small kill zone on this animal—you need to aim low on the body, right behind the shoulder.

I've also had good luck with coyotes while turkey hunting. I set my turkey decoys out and proceed to call turkeys. I've had several coyotes charge my decoys and have been able to take them with a shotgun. This makes for a very exciting hunt."

—Sheila Dyer, certified bow technican and certified shooting instructor, The Old Trading Post, Belfast, Ohio

of day when coyotes are least likely to be there, such as midday. After the coyotes have had a long time to get used to the idea of these free eats, such as a couple of weeks, find yourself a good roost, about 30 or 40 yards away, and lie in wait to hunt or photograph them.

Can you hunt coyotes from the air?

Though both expensive and dangerous, this is actually one of the best ways to make sure you are killing coyotes only, when trying to reduce the coyote population or remove problem coyotes. With special permits, government agents aerial-hunt coyotes from both low-flying, small fixed-wing planes and helicopters, with the latter being favored for less open, forested, and rugged terrain. When there is snow on the ground, the resulting tracks make coyotes easier to locate, and ground crews may also be used to locate coyotes ahead of the aircraft, with howlers. Coyotes in many areas, of course, have learned what that buzzing or clattering in the sky means, and head for dense cover at the first sound of it.

Is it true that hunting with dogs is one of the best ways to hunt coyotes?

If by best one means most effective, that is definitely true. Trained dogs can be used in a number of ways to increase your chances of killing coyotes. Some of those ways are thoroughly condemned by people who feel that they are unfair and inhumane.

One of the oldest methods of hunting with dogs is using what are called sight hounds to run down coyotes. These are the speedsters of the dog world, such as greyhounds and salukis, and some sight hounds, such as Irish or Russian wolfhounds, are huge as well. The dogs are transported in cages in the back of a pickup or other truck, and released when a coyote is sighted. Dogs like these can outrun even the swift coyote in short distances, and not long after they are released the coyote is dead on the ground. This method has a high chance of success, although feeding and keeping a pack of dogs like this is a high price for it.

A more recent approach is to use medium-sized dogs to draw coyotes into gun range in breeding season. Coyotes are extra-territorial when they have an active den, and when a dog enters the area they will pursue it vigorously. The dogs involved here are trained to lead the coyotes back to the hunter. Breeds favored for this method include the Mountain Cur, which is big enough to defend itself against coyotes.

The most intensely contested method of hunting coyotes with dogs involves attracting coyotes to an area, often with deer carcasses, and then setting packs of dogs on them to run them to exhaustion. The dogs wear GPS equipment, so their handlers can locate them when a fresh shift of chasers is needed. Practitioners of this approach are believed to sometimes intentionally wound (not kill!) the weary coyote to be sure it has a chance to be torn to pieces by the killer dogs that are eventually released. Critics of this method call it legalized dogfighting.

What about den hunting? Exactly what is that?

Den hunting has long been practiced by government predator control agents and bounty hunters, and is still often an effective way to reduce livestock or pet depredation in a particular area, because a pair of adults feeding pups are among the most active of predators. However den hunting is not something people who love baby animals want to hear about.

First the den is located using howlers, tracking, trail-following, dogs, and sometimes aircraft, plus the clues to den location described earlier. The adults are killed upon approach, if possible, with a rifle or shotgun. Then the pups are destroyed in the den with a carbon monoxide bomb or gas cartridge, or dug out and destroyed. They may also be brought out with a length of barbed wire called a feeler, which is fed into the den until it snags them and brings them to the surface. Once the pups are gone, so is the constant need to feed them, so predation in the immediate area usually diminishes.

CHAPTER 8

a few last facts about coyotes

Here you will find a variety of additional interesting information about coyotes, including how to watch and photograph them; the pros and cons of coyotes as pets; diseases that can or cannot be caught from them; how long coyotes live; and what forces they normally succumb to.

COYOTE-WATCHING

If you would like to do some coyote-watching, either for the simple pleasure of observing such a dramatic form of wildlife, or to spy on them for some less innocuous future purpose, you have probably figured out by now that this, like just about every attempt to get a closer look at coyotes, is not easy. Coyotes owe much of their success to their ability to stay largely invisible. Their keen senses make sneaking up on them difficult, and they are extra wary in places where they have been hunted or trapped. It now may be easier to get a glimpse of them in some suburban areas, but even there, they are most likely to disappear before you have satisfied your curiosity. The following tips will be helpful in coyote reconnaissance.

A Colorado coyote in a field of wildflowers.
Courtesy of Colorado Division of Wildlife (Michael Seraphin)

When are you most likely to see them?

Since coyotes are night hunters for the most part, and few of us are out wandering around looking for animals then, the most likely times for coyote sightings in rural areas are right after sunset and right before sunrise. They sometimes may be seen hunting in the daytime in rural areas where they feel safe, or when they are pressed to feed a den full of pups. Daytime sightings are less common but have been reported in suburbs and urban areas.

As for time of year, coyotes are most often seen by us during their breeding season, from January to March, when they are out and about seeking mates. In early fall, when the parents are teaching the pups to hunt, you may be lucky enough to catch sight of a family group. Coyotes are easiest to spot during the winter months in northern areas, because there is less vegetation to hide them, and they will stand out against any snow. And snow will also mean tracks to help one figure out the most likely viewing areas. When snow has covered the ground for several days, coyotes may hunt in late morning or midday.

A clue to coyote presence

The distress calls of small birds or squirrels, which may have spotted a coyote you haven't yet, are sometimes a clue to coyote presence. Sightings of magpies or ravens may be another tipoff, since both of these scavengers like to follow coyotes in hopes of some leftovers.

The best places to look

The best places to look for coyotes are open fields, pastures, and other grassy areas, the edges of woods, and ponds or other water sources that are hidden from view. Coyotes also follow game trails, often using them at the same time each day, so a spot overlooking one is a good choice, as is a place with a view of a canyon or ravine.

To up the odds

To increase the odds of a successful stakeout if you live in a rural area, you can set up some bait, such as the carcass of a dead calf or sheep or a road-killed deer. Scent stations are good attractants too.

Keeping yourself out of sight

Keeping yourself out of sight is key, of course, as you await a coyote. This means hiding in some kind of good cover, not too close (not closer than 100 feet) to your lure or observation area. If you plan to use a blind, it must be set up well ahead of when you intend to use it—ideally as much as a whole month ahead, so the coyotes have time to get used to it. Remember, coyotes do not like new things in their environment. If you intend to use a commercially made blind, even in camouflage colors, it is always a good idea to further camouflage it with some of the surrounding vegetation.

Once you are at your observation post, you must stay as STILL as possible. Bring along the means to take care of potty needs with as little movement and commotion as possible, and to dispose of urine and feces in a way that will not pollute the surroundings or give scent warnings of your presence. And ideally, you should also treat yourself and your clothing to minimize human scent. Bring something comfortable to sit on, too, to help you be as motionless as you can manage.

Eyeball assistance

A good pair of binoculars, such as 10 x 40 or 8 x 42, is often called for, since we mostly are viewing coyotes at some distance. You might also consider a spotting scope, for observations at greater distances, or even night-vision binoculars. A third-generation night-vision scope, if you can afford it, is ideal. Often a good pair of eyes scanning in all directions, followed by a more close-up view with binoculars or a spotting scope, is the most effective approach for locating coyotes.

Stay where you are!

If you do see coyotes, stay where you are—do not approach them. Especially do not approach a den, should you happen to stumble upon one. Not only are coyotes very protective of their den, if there is human disturbance even as far as a quarter of a mile from it, they will move it. No matter how much you like coyotes, don't try to make friends with them. Even when this is not dangerous to you, it will ultimately be dangerous to them, if it lessens the wariness that makes peaceful coexistence with us possible. "Even if you love coyotes, never let them know it," as one expert said.

CAPTURING THE MOMENT

If you are one of the many people who like to capture and preserve wildlife sightings for later enjoyment, to brag a little to others, or add a touch of the wild to the walls of your home or the desktop of your computer, you will of course want to try photographing coyotes. Getting good shots of wary wildlife is rarely easy, and the coyote is one of the top challenges. Longtime pro Steve Maslowski, of Maslowski Wildlife Productions of Cincinnati, offers some pointers here:

- "Since it is usually hard to get close to coyotes, long lenses are the rule for photographing them. How long? Anything less than 600 mm is going to handicap you.
- "A few extra million pixels in a camera sensor can help, too. A camera with a high megapixel count (say, greater than ten) allows more enlargement so you can crop in closer to that animal filling out only a disappointingly small part of the frame.
- "Along with a potent camera, a predator call can help. Predator calls imitate the squalls of an injured prey species such as a rabbit. Available from big outdoor sporting stores, predator calls come in mouth-blown and electronic versions. Predator calls are highly alluring to a hungry coyote . . . but you still need skill and luck to predict his approach, and to overcome problems with wind that carries your scent, and line-of-sight issues.
- "Having said all this, probably the greatest trick to photographing coyotes is to go to certain national parks. In my own limited experience, in Yellowstone and Grand Canyon coyotes are a fairly common sight, and for the most part the coyotes there regard photographers as just part of the scenery. There are surely many other parks where they are equally as common and accommodating . . . and here's hoping you find one soon."

Setting up a bait station, as described on page 160, is another good way to lure coyotes into view (remember that you **never** want to feed coyotes in or near a residential area!). Using a blind, as outlined earlier, is a traditional way to get close to your quarry without alarming it. Photographers who use tripods in the open sometimes use camouflage netting to cover both the camera and its support.

In the backyard or at a national park, or wherever you may be trying to get a good shot of a coyote, never get too close, even if the coyote in your viewfinder would let you.

Catching them unawares (spyware)

One of the surer ways of getting some kind of picture of coyotes—it may not be the clearest or best-composed one—is to use one of the many kinds of game cameras available today. These are cameras that you set up outdoors in a likely spot, and then let the coyotes themselves take their picture. Cameras like this are activated by motion or body heat and are often set up next to a likely trail, gathering spot, or an attractant such as a scent station or serve-yourself buffet for the animal in question.

As to choice of models, the more megapixels, the better the picture quality will be. Experienced game-camera users say that the infrared type of remote camera is best because it doesn't create a highly visible flash that spooks your four-legged visitors.

Arranging the camera so that it sits about ten yards away from the anticipated subject(s) is best—too close and all you get is a whiteout. If you are using an LED (color) model, always point the camera either north or south, to avoid sun flares in the camera lens.

WHAT ABOUT HAVING A COYOTE AS A PET?

If you've ever seen a picture of a litter of coyote pups, or been lucky enough to see the real thing, you would have to have a heart of stone to not be charmed. Even the most hardened coyote exterminators, digging up a den, have been moved. Coyote pups, with their short brown or gray fur, short ears and muzzles, and bright little eyes, have the helpless-looking, pudgy appeal of all baby animals, and then some. Thus, many an attempt has begun to keep one of these darling little creatures as a pet.

From a very early age, as with dog pups, some coyote pups are simply friendlier and more outgoing than others, so that would probably be the one you picked (assuming that you have gotten whatever permits may be necessary to keep a wild animal in your state). And in the beginning, this would be a win-win proposition. Your efforts to hand- or bottle-feed the little creature would be rewarded with an animal that waits for your visits, likes to be petted and scratched, wags its little tail, licks you, and snuggles up to you at night. And it is every bit as playful and lovable as any other puppy. It would probably become best friends with any dog you might have or, for that matter, cat. The popular book and Web site *The Daily Coyote* by author and photographer Shreve Stockton has eloquently recorded all of these delights, as do some of the chapters of eastern coyote researcher Jonathan Way's *Suburban Howls.*

True, as it grows, a coyote pup may play a little rougher than a dog pup, and you may or may not be successful at housebreaking it. Shreve Stockton and others have reported success at housebreaking (her pet coyote, Charlie, came up with the idea of peeing down the bathtub drain, which pleased both of them). Or you may never be able to have a house that contains an inside coyote and also has unsullied rugs and floors.

But as time passes, that will not be your biggest problem. Gradually, it will become clear that the very things that make coyotes magnificently designed for life in the wild, make them poorly suited to be pets. As Hope Ryden, one of their most devoted defenders, once pointed out, in a coyote you have a "dog" with heightened sensitivity, lightning responses, and a highly independent spirit—in short, a doglike creature without the 10,000 years of domestication that have made our domestic dogs what they are, for better and worse. Or as Frank Dobie, another highly respected student of coyotes noted, "The Russians have a proverb. You may feed a wolf as much as you like, but he will always glance at the forest."

As the months go by, your adorable little ball of fluff will get larger and taller and rangier, and its behavior will change, too. It may become less friendly and want to go off on its own for part of the day and night. Then the day will come when it suddenly growls at you, or challenges you. Even Shreve Stockton's cherished Charlie reached this point, and she had to keep a deer antler handy to defend herself. Suddenly, she never knew when he might snarl at her, jump on her, or even bite her. To work her way back to a point of greater control, she had to buy and digest a book on the "pack leader" style of dog training. As a pet coyote grows older, being the dominant, assertive leader of the pack is the only way you will keep the upper hand.

Even so, you will always be at best a pet coyote's keeper, not its master. Coyotes have too much self-respect to be subservient to people. You may be able to train them to a point, but they are not going to indulge your every whim, or do tricks for you. As one coyote pet owner pointed out, "you don't discipline them, you learn to cooperate with them."

Pet coyotes are generally one-man or -woman pets, as well. They may not mind your visitors when they are pups, but as they age they are generally nervous and shy, fearful of strangers. They are also very uneasy about even the smallest change in their environment—whether the rooms of your home or their pen outdoors. Both of these things are natural reactions for what is, after all, a wild animal, but it does not add to their appeal as pets. Hand-raised wolves, by and large, are much friendlier. Is a captive coyote cowering in the corner of his pen really what we are after?

All of the above (plus frequently a growing interest on their part in things like their owner's chicken coop or rabbit pen) often add up to owners falling out of love with their coyote pets after they grow up, or at least deciding they can no longer keep them. The outlook then is bleak, as these animals are too used to humans by this time to make it in the wild, and too wild to make good pets for anyone else. So they are usually abandoned somewhere, where they end up in trouble (such as raiding livestock, or begging for handouts along the road), and are soon shot or trapped. Or they may be dumped at an animal shelter that is not equipped to deal with them either. A few lucky ones may end up at a zoo able to take them in.

The bottom line here?

As Hope Ryden also noted, "a few rare individuals find life with such a high-strung creature rewarding. Do not assume this burden [note the word "burden"] without being willing to adapt to the animal's lifestyle." Shreve Stockton, who loved her Charlie beyond all imagining, did just that. But bear in mind that she lived in a part of the country that was not highly populated (where there was less chance of her beloved pet being blown away by someone who assumed it was just another of those "awful coyotes"). Stockton also was willing to mold her entire lifestyle around her pet coyote, give it endless time and affection (even at the risk of losing human relationships over it), and eventually, bought it a half-acre yard with a very expensive coyote-proof fence around it. Not everyone is willing to go this far, even for what is a very attractive-looking animal, as her photographs attest.

DO COYOTES CARRY RABIES, OR ANY OTHER DISEASE MY PETS OR I COULD CATCH FROM THEM?

For centuries, rabies has been feared—with good reason—as a disease capable of producing, for man or animal, very nasty symptoms and, if left to run its course, an inevitable and agonizing death.

There are several thousand cases of rabies a year in this country, in wild and domestic animals (even cattle!), but at this time—and in recent history—there have not been a great number of rabid coyotes. There are far more rabid raccoons, skunks, foxes, and bats. In fact, some scientists feel that by preying on animals like these (plus feral dogs and cats), coyotes actually cut down on the development and spread of rabies.

But coyotes do get rabies, usually from other wild animals, and they can spread it to humans or pets. Studies have shown that coyotes who act aggressively toward humans, or bite people or pets, are much more likely to be carrying this disease. In 2004, for example, a rabid coyote attacked a man who was mowing his lawn in Vancouver, and in 2006 one attacked a Pennsylvania family's dogs and tried to force its way into their home, until a family member killed it with a shotgun. Texas had a number of rabies outbreaks in coyotes in the late 20th century, and there is some concern about the potential of the large new populations of urban coyotes for contracting and then spreading the disease.

The relationship between aggression and rabies is the reason authorities usually insist that a person be treated for possible rabies exposure when a human and/or their pet has been attacked by a coyote. Due to recent advances in the science of immunization, this is not as traumatic a course of treatment as it once was.

Rabies is a disease of the central nervous system that causes inflammation of the brain. It is transmitted when saliva from an infected animal comes in contact with broken skin, or more commonly, by a bite. There are two different forms of the disease. One causes animals to act uncharacteristically bold, biting and snapping at anything in their path. The other causes them to act unnaturally tame and friendly. Either of these behaviors is good reason to stay well away from an animal and report it to animal control or other authorities as soon as possible.

If you or your pet has been bitten or scratched by a coyote, scrub the area(s) well with soap and water—five minutes is recommended—and rinse well. Then apply iodine and call the health department and/or the vet. Do not delay seeking medical help and advice. Treatments for rabies will not be successful unless begun before the onset of symptoms.

After an attack, your state or local authorities will attempt to find and euthanize the coyote in question. If you happen to have somehow captured it alive, keep it confined and well away from people and other animals until the authorities arrive. Positive identification of rabies usually involves an examination of the animal's brain.

Making sure that your own dogs and cats are all vaccinated regularly for rabies is an important preventive measure here.

Distemper

Distemper is another serious canine disease coyotes can contract. Distemper has been spread by domestic dogs to various forms of wildlife in the United States, and they can spread it back to pet dogs. Distemper takes a high toll on coyote pups especially. Animals can catch it from direct contact with an infected animal, or contact with urine, droppings, or mucus from sick animals. Canines with distemper are listless, run a fever, and have runny noses and eyes. They may vomit, cough, or have diarrhea, and they usually lose weight. Infected animals sometimes have nervous symptoms as well, such as involuntary movements of their jaws, and the whites of their eyes may become discolored. Coyotes with distemper are usually destroyed to prevent spread of the disease.

Parvovirus

Parvovirus is another, and newer, virus disease that coyotes can contract as well as dogs. It is not only deadly but highly contagious, and is spread by contact with the feces of infected animals. Puppies are the most susceptible. Animals with "parvo" are lethargic, lose their appetite, vomit, and have bloody diarrhea. The latter is often so severe that they die of dehydration within 72 hours or less.

Rabies, distemper, and parvovirus are three good reasons to keep your dogs away from wild coyotes and make sure they have regular inoculations for these diseases. Pet coyotes should also be immunized for the "big three."

Mange

A parasite coyotes are particularly susceptible to is the mange mite, which leads to a condition called sarcoptic mange. A tiny mite, something like the chigger that can make human lives miserable, burrows into the skin of the coyote and lays its eggs. As the mites proliferate and spread from the legs and hips to the flanks and head, the coyote is so busy scratching himself against anything handy that it may forget to hunt and eat. It is

A four-month-old coyote pup from Connecticut, in the early stages of mange. Photo by Patricia Osborne

soon a balding mass of open sores, and within a couple of months, emaciated, and then dead. Who would think that this miniscule relative of ticks, spiders, and scorpions could do this? But books by coyote experts have plenty of gruesome photos of mange-killed coyotes to prove it. Mange tends to strike coyotes periodically, and at times as much as 60% or more of a population can be wiped out by it.

If you see a really ugly and ratty-looking coyote, it probably has mange. Mange is contagious and can be transmitted to pets by coyotes, but it rarely is. Probably because dogs are unlikely to snuggle up to a coyote, and cats who meet a mangy coyote may not survive to catch it.

Heartworm

Heartworm is not something your dog can catch directly from a coyote, but it may indirectly. A number of urban coyotes, especially, now have heartworm disease, scientists have discovered, and this can create a nearby "reservoir" of the disease, from which it can be spread via mosquitoes to your pooch. Make sure your dog is given heartworm preventives regularly.

Other diseases

The chance of you or your pet catching any of these diseases from a coyote is remote, but just to complete the list here: Dog ticks or wood ticks on coyotes can transmit Rocky Mountain Spotted Fever, and coyotes can also, with the help of their fleas, carry bubonic plague, tularemia, and equine encephalitis. Coyotes can also catch and carry canine hepatitis, and they have a number of parasites, including tapeworm, hookworm, and lungworm. Some of these can wreak havoc in humans, affecting parts of the body they never would in their normal hosts, so avoid handling coyote scat with your bare hands.

To protect your health and your pets

Report immediately any coyote that is acting abnormally—that seems slow moving, disoriented, ill, aggressive, or unafraid (which is not good even if it is *not* sick!).

If you have to handle a dead coyote, use rubber gloves or a shovel, and wash your hands well afterward.

HOW LONG DO COYOTES LIVE?

The answer to this question is "about as long as the average dog," but the real question here is not so much how long *can* they live, but how long do they normally live? Coyotes for the most part have a high-risk lifestyle.

As senior-citizen coyotes in captivity have shown (the record holder was a coyote at the National Zoological Park in Washington, D.C.), coyotes can live for up to 18 years. And a few 13- or 14-year-olds have been found in samplings of wild coyote populations in places like Colorado, Texas, and Alberta. Of 1,558 coyotes killed in Minnesota from 1968 to 1976, the oldest was 11 years. In a Texas study, one of 67 coyotes caught was over 8 years of age, and of 321 sampled in Nebraska, the oldest was 9 years of age. But as coyote guru Stan Gehrt has noted, in the wild most coyotes die before they reach their third birthday.

About 50 to 70% of coyote pups die in their first year, and the yearly death rate for adult coyotes, who are more experienced and socially secure, is 30 to 40%. If pups survive the diseases of puphood, and escape the attentions of hawks, eagles, poisonous snakes, bobcats, and bears, they have a very good chance of being killed as they head off in the fall to seek their own territories and mates on a journey that may be hundreds of miles, and is full of roads and highways that are high-risk travel lanes for exuberant youngsters.

Adult coyotes can be killed by wolves, large dogs, and cougars, and some do die of starvation and diseases like mange. But for the most part their causes of death are human related. Death by speeding vehicles, as noted above, is one big reason. In cities and suburbs this can amount for as much as 70% of coyote deaths every year. Other main ways coyotes meet an untimely end are hunting, trapping, and poisoning. Coyotes are no longer poisoned on the scale they once were, but it still happens.

Photo by Steve and Dave Maslowski

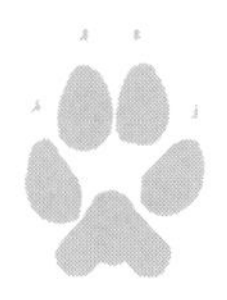

parting thoughts

A CRYSTAL BALL FOR COYOTES?

What does the future hold, when it comes to coyotes? For many years, humans spent far more time and money trying to kill coyotes than studying or attempting to understand them. Now at last coyotes are stars and darlings of the research agenda and funding sources, so before too much longer we should know things we don't know now that will smooth and improve our ways of coexisting with them.

It does not take a scientist, however, to see what is clear after reading all of the books, articles, and Web site information I have assimilated in the course of researching this book. Everything depends on where coyotes go from here in their increasing boldness—though in any case it is hard to not conclude that attacks on pets and for that matter people are only going to increase in the days ahead. To quote Stan Gehrt of The Ohio State University one more time: "Coyotes have to be taught that humans are boss. We believe it's easy to do with animals that are just starting to test the waters. But with other coyotes that have been habituated for decades, we just don't know. With each generation, they become a little more familiar with people and the landscape of people. Where that ends, nobody knows. Coyotes are watching and learning from us, we influence their behavior, and it will be our actions that determine what the future holds for our new neighbors."

To forestall a futuristic fantasy–horror film called *The Day (Night?) of the Coyote,* we need to find good ways to stem the tide.

COEXISTING WITH COYOTES

Since it would be close to impossible to eliminate coyotes from all of the ground they have taken, even if we wanted to, what we need to do is find ways to coexist with them. This means seeking and putting into effect measures that will minimize the harm they are able to do, so that we can sit back and enjoy the touch of the wild they bring to even the most urban surroundings. Several of the chapters of this book outline suggestions meant to accomplish this, especially Chapter 6.

One might say that we are living today with a new wolf, a smaller, more subtle and streamlined creature that suits today's surroundings, and that in many ways mirrors us, more than we would ever admit to. This new wolf—the coyote—is an extraordinarily adaptable creature, but if we are the super-intelligent, super-adaptable creatures *we* are alleged to be, we will somehow find ways to live in harmony with it.

RECOMMENDED READING

Eastern Coyote: The Story of Its Success, by Gerry Parker. Halifax, Nova Scotia: Nimbus Publishing, 1995.

The Voice of the Coyote, Second Edition, by J. Frank Dobie. Original edition 1949 by Curtis Publishing, new edition 2006 by Bison Books of Lincoln, NE (The University of Nebraska).

Coyote, by Wyman Meinzer. Lubbock, TX: Texas Tech University Press, 1995.

The World of the Coyote, by Wayne Grady. San Francisco: Sierra Club Books, 1994.

The Daily Coyote: A Story of Love, Survival, and Trust in the Wilds of Wyoming, by Shreve Stockton. NYC: Simon and Schuster, 2009.

Suburban Howls: Tracking the Eastern Coyote in Urban Massachusetts, by Jonathan G. Way, PhD. Indianapolis, IN: Dog Ear Publishing, 2007.

God's Dog: The North American Coyote, by Hope Ryden. An Authors Guild Backinprint.com edition published by iUniverse; originally published by Coward McCann in 1975.

The Coyote: Defiant Songdog of the West, by Francois Leydet and Lewis E. Jones (illustrator). Norman, OK: University of Oklahoma Press, 1977; revised and updated edition 1988.

Coyote: Seeking the Hunter in Our Midst, by Catherine Reid. Boston: Mariner Books (Houghton Mifflin), 2005.

Coyote at the Kitchen Door: Living with Wildlife in Suburbia, by Stephen DeStephano. Boston: Harvard University Press, 2010.

Coyotes: Behavior and Management, edited by Marc Bekoff. Caldwell, NJ: The Blackburn Press, 1978.

Coyotes: Predators and Survivors, by Charles L. Cadieux. Washington, D.C.: Stone Wall Press, 1983.

Solving Coyote Problems: How to Outsmart North America's Most Persistent Predator, by John Trout, Jr. Guilford, CT: The Lyons Press, 2001.

Nuisance Animals: Backyard Pests to Free-Roaming Killers, by John Trout, Jr. Tennyson, IN: Midwest Publishing, 1997.

Living with Coyotes: Managing Predators Humanely Using Food Aversion Conditioning, by Stuart R. Ellins. Austin, TX: University of Texas Press, 2005.

"Urban Coyote Ecology and Mangement: The Cook County, Illinois, Coyote Project," by Stanley D. Gehrt. Bulletin 929, The Ohio State University Extension Service.

Mammal Tracks & Sign: A Guide to North American Species, by Mark Elbroch. Mechanicsburg, PA: Stackpole Books, 2003.

A Guide to Animal Tracking and Behavior, by Don and Lillian Stokes. Boston: Little, Brown, and Company, 1986.

The Complete Tracker: Tracks, Signs, and Habits of North American Wildlife, by Len McDougall. Guilford, CN: The Lyons Press, 1997.

Coyote Hunting: The Eastern Coyote, by Tom Bechdel. Terra Alta, WV: Headline Books, 2006.

The Coyote Hunter: A Complete Guide to Tactics, Equipment, and Techniques for Hunting North America's Perfect Predator, by Don Laubach, Merv Griswold, and Mark Henckel. Gardiner, MT: Don Laubach, 2000.

Navaho Coyote Tales, collected by William Morgan. Santa Fe: Ancient City Press, 1988.

INDEX

DEAR CUSTOMERS AND FRIENDS,

SUPPORTING YOUR INTEREST IN OUTDOOR ADVENTURE, travel, and an active lifestyle is central to our operations, from the authors we choose to the locations we detail to the way we design our books. Menasha Ridge Press was incorporated in 1982 by a group of veteran outdoorsmen and professional outfitters. For many years now, we've specialized in creating books that benefit the outdoors enthusiast.

Almost immediately, Menasha Ridge Press earned a reputation for revolutionizing outdoors- and travel-guidebook publishing. For such activities as canoeing, kayaking, hiking, backpacking, and mountain biking, we established new standards of quality that transformed the whole genre, resulting in outdoor-recreation guides of great sophistication and solid content. Menasha Ridge continues to be outdoor publishing's greatest innovator.

The folks at Menasha Ridge Press are as at home on a white-water river or mountain trail as they are editing a manuscript. The books we build for you are the best they can be, because we're responding to your needs. Plus, we use and depend on them ourselves.

We look forward to seeing you on the river or the trail. If you'd like to contact us directly, join in at www.trekalong.com or visit us at www.menasharidge.com. We thank you for your interest in our books and the natural world around us all.

SAFE TRAVELS,

Bob Sehlinger

BOB SEHLINGER
PUBLISHER